THE PROBLEM OF RATIONALITY IN SCIENCE AND ITS PHILOSOPHY

BOSTON STUDIES IN THE PHILOSOPHY OF SCIENCE

VOLUME 160

THE PROBLEM OF RATIONALITY IN SCIENCE AND ITS PHILOSOPHY

On Popper vs. Polanyi
The Polish Conferences 1988–89

Edited by
JÓZEF MISIEK
Jagiellonian University, Kraków, Poland

KLUWER ACADEMIC PUBLISHERS
DORDRECHT / BOSTON / LONDON

Library of Congress Cataloging in Publication Data

The Problem of rationality in science and its philosophy : on Poper vs. Polanyi, the Polish conferences 1988-89 / edited by Józef Misiek.
p. cm. -- (Boston studies in the philosophy of science ; v. 160)
Includes index.
ISBN 0-7923-2925-2(acid-free paper)
1. Science--Philosophy--Congresses. 2. Rationalism--Congresses. 3. Popper, Karl Raimund, Sir, 1902- --Congresses. 4. Polanyi, Michael, 1891- --Congresses. I. Misiek, Józef. II. Series.
Q174.B67 vol. 160
501--dc20 94-17897

ISBN 0-7923-2925-2

Published by Kluwer Academic Publishers,
P.O. Box 17, 3300 AA Dordrecht, The Netherlands.

Kluwer Academic Publishers incorporates the publishing
programmes of D. Reidel, Martinus Nijhoff,
Dr W. Junk and MTP Press.

Sold and distributed in the U.S.A. and Canada
by Kluwer Academic Publishers,
101 Philip Drive, Norwell, MA 02061, U.S.A.

In all other countries, sold and distributed
by Kluwer Academic Publishers Group,
P.O. Box 322, 3300 AH Dordrecht, The Netherlands.

Printed on acid-free paper

TABLE OF CONTENTS

JÓZEF MISIEK

PREFACE

Rationality of science was the topic of two conferences (held in 1988 and 1989) organized by the Department of Philosophy of Science, Institute of Philosophy, Jagiellonian University. Both conferences included a small group of invited speakers. This book contains a selection of papers presented there. It is intended mainly for specialists in the philosophy of science and scientists interested in philosophy. Students and especially postgraduate students would also benefit from reading it.

The first conference, 'Popper, Polanyi and the Notion of Rationality', was held from 1 to 5 October 1988 in Janowice. The second conference, 'The Aim and Rationality of Science', was held in Cracow at the Jagiellonian University, from 4–10 June 1989.

The topics of both conferences were inspired by our late friend Dr. Tomasz Kocowski, who many years earlier invited me and my colleagues from the Department to participate in research concerning the problem of creativity, and serve him and other psychologists as methodological advisors. Personal contacts with this intelligent and inquisitive man helped us to realize that we could not fulfill our task while adhering to the received view in the philosophy of science. This experience helped us to see science not only as scientific knowledge but also as a process of research. We then turned our attention to Michael Polanyi, who seemed to provide the philosophy we were looking for.

Organizing conferences in a communist country was a special task which would have been impossible to perform without the friendly cooperation of my colleagues from the Department. I want to thank especially Doc. Dr Hab. Zdzisława Piątek and Dr. Janusz Płazowski.

The first conference was supported by the Polish Ministry of Higher Education. The second was sponsored by the Latsis Foundation, due to the kind recommendation of Prof. John Watkins from the London School of Economics. I would like to express my gratitude to him and both institutions.

It is a pleasure to thank the general editor of the *Boston Studies*, Prof. Robert S. Cohen, for his encouragement, for securing anonymous referees of the manuscripts, and for substantial contribution to the editing of manuscripts. I am much in debt to Annie Kuipers of Kluwer Academic Publishers for her

J. Misiek (ed.), The Problem of Rationality in Science and its Philosophy, vii–viii.

guidance of and patience with the inexperiential editor.

Dr. Marek Suwara provided much needed linguistic help during the editorial process, and Dr. Janusz Płazowski was very helpful in dealing with computer problems.

Last but not least, my special thanks to all participants of both conferences for bringing with them their own original ideas, discussing problems in a friendly but demanding atmosphere, and postponing publication of their contributions until the date of printing of the present volume.

Kraków
November 1993 Józef Misiek

JÓZEF MISIEK

INTRODUCTION

The papers presented here come from two conferences, both concerning the problem of rationality. The choice of this topic reflects the general opinion among specialists that rationality is a crucial problem in the philosophy of science. What is unusual is the fact that the first conference, 'Popper, Polanyi and the Notion of Rationality' was intended as a confrontation between practitioners of the standard approach to philosophy of science, personalized by the name of Karl Popper, and the unorthodox approach inspired by Michael Polanyi.

The second conference gathered many leading philosophers from abroad. It can be seen as a review of leading tendencies concerning the problem of rationality and related problems in the philosophy of science.

Most of the papers in this volume may be divided into two types, named after the philosophies that inspired the views: the Popperian view and the Polanyi-inspired position. One should stress that none of the papers represents a dogmatic attitude, but rather attempts some kind of improvement of the particular position adopted.

Let us start our survey from the Polanyi-inspired papers. Prof. Cattani, attempting a better understanding of Polanyi's philosophy, puts it in the broader context of the rhetorical tradition in philosophy taken in its ancient rational dimension. Analysis of the Polanyi's notion of rationality relates well with remarks concerning the political implication of this philosophy. Prof. Scott uses a strategy similar to Cattani's. He analyzes Polanyi's philosophy in the broader context of combat, against both scientism and totalitarian attitudes in social philosophy. As a result we obtain a very clear picture of Polanyi's philosophy.

Prof. Życiński attempts to show that the alleged irrationalism of Michael Polanyi's philosophy is a misunderstanding. He even claims that between the Scylla of dogmatic rationalism of positivist provenience, and the Charybdis of relativism, one has to work out a new epistemology stemming from both Popper's and Polanyi's contributions to philosophy.

A similar conclusion follows from the paper 'Personal Rationality' written by the present writer. This paper offers comparisons between Popper and Polanyi and attempts to show that some of Popper's formulations resemble

J. Misiek (ed.), The Problem of Rationality in Science and its Philosophy, ix–xii.

Polanyi's theses. The second paper by the same author, 'Assessment of Theories', pertains to the Polanyi-oriented analysis of the assessment of special relativity. In view of historical facts, the paper proposes to distinguish notions of egalitarian and elite rationalities.

The last paper to show some influence of Polanyi is by Dr. Werszowiec Płazowski. He adopts a very broad context to discuss the problem of rationality, that of the whole culture. Within such a context Płazowski tries to overcome the Popperian dichotomy of science and metaphysics, and comes to the conclusion that theory without facts is empty and theory without metaphysics is dead.

The more numerous group of papers presents different versions of Popper's inspiration. Prof. Agassi puts the problem of rationality of knowledge *vis-a-vis* rationality of social order. He makes clear that both kinds of rationality are necessary for mankind to survive. His conclusion is that no theory of rationality offers a proper solution to this problem.

Prof. D'Agostino presents an interesting counter-proposal to Prof. Watkins's solution of the problem of rationality.[1] According to him rationality of science should not be connected with the aim of science, as Watkins claims, but rather with commitment to some specific rules of behavior: rationality is not teleological but deontological. It is worth noting that both ideas involve personal aspects of science, which seems to be against the Popperian tradition but perfectly matches the Polanyian one.

Another criticism of Watkins[2] comes from Prof. Krajewski. Krajewski accepts the essentials of Watkins's proposal but finds it necessary to modify the particular set of claims accepted by his opponent. One such modification follows from the postulate of realism. A similar position is taken by Prof. Boyer who argues that some of Watkins's aims of science should be abandoned unless realism is not postulated by the second philosopher.

There are papers which are influenced by the Popperian school. Dr. Grabińska finds difficulties in the Lakatos-Zahar reconstruction of two research programs pertaining to the problem of the electrodynamics of moving bodies. One of them, Einsteinian, is well known. The second, developed by H.E. Ives, is known only by specialists. The paper makes a direct assault on the Lakatosian notion of rationality. Prof. Piątek's paper has a similar orientation, she focuses on the problem of rationality in Darwin's theory in light of Popper's thesis that it is not a theory but only a metaphysical research program. Piątek finds interesting arguments against Popper's thesis.

Prof. Watkins's contribution attempts to make the notion of theory more precise in order to clarify the concept of scientific progress. The latter concept has direct bearing on the concept of rationality. Watkins assumes understanding of a theory as axiomatised (even formalized). The problem is solved in terms of Weisberg's requirement.

Dr. Zabierowski analyses Popper's dichotomy subjectivism-objectivism and confronts it with contemporary science. He finds that the dichotomy does not hold and a third element must be introduced: anthropism. There is also a paper criticizing the sociobiological theory of knowledge, written by Prof. Życiński. The author shows that the sociobiological theory of knowledge in Ruse's version is incompatible with rationality.

Other papers can not be assigned among either the first or the second group. Each of them makes a separate group. Prof. Armstrong's proposal is, in order to make induction rational, to count it as a particular species of abduction pertaining to properties of things that are universal. Prof. Churchland starts from a nuerocomputational perspective (a branch of artificial intelligence adopting parallel processing). The paper offers an interesting account of the concept of explanation. But it gives more. It also offers a general concept of rationality quite close to that of Polanyi. Even the concept of tacit knowledge is included. It is modeled by a configuration of synoptic weights in a neural network, a configuration never defined by the research worker but acquired by the network during the process of training. Prof. Przełęcki's paper presents a very able and crystal clear defense of the received view. Going against the main current, he criticizes some objections raised against the received view pertaining to the problem of interpretation of the theory's language. Prof. Musgrave is engaged with the problem of scientific realism. Adopting a metaphysical approach, he offers able arguments against contemporary objections to realism. In particular he criticizes those presented by Nancy Cartwright in her book *How the Laws of Physics Lie.* Prof. Tempczyk applies recent discoveries in the theory of dynamical systems in order to show that classical mechanics is far from deterministic.

It is interesting to notice the substantial overlap of ideas among many papers. Thus, for example, the main concept discussed in Popperian papers is the notion of the aim of science. Such an idea clearly goes beyond the limits of objective knowledge and introduces a personal dimension to the philosophy of science. Thus the Popperian school takes a step toward the Polanyian tradition.

Another interesting feature of many papers is more or less explicit rejection of the sentential conception of knowledge, rejection of logic as the sole legitimate tool of philosophical analysis, emphasis on understanding knowledge not just its syntactic features. The most striking, however, is more or less tacit acceptance of the concept of tacit knowledge. It can be clearly seen in Popper-oriented papers but is most perspicuous in Churchland's approach.

Another interesting phenomenon exhibited by most papers in the volume is the strong attack, coming from different positions, on the received view. This attack differs from the well-known 'revolutionary' philosophy of Kuhn and Feyerabend in that it offers specific proposals for how to develop philosophy of science, proposals much more concrete than 'anything goes'. It is reassuring to see that the combat with the 'received view' and the neopositivist heritage is not at all nihilistic as it used to be. On the other hand, the received view is still alive as some papers show. The confrontation between these different approaches results in an intellectual ferment which can be considered fruitful for philosophy of science. All philosophers of science who are open-minded will find many stimulating, even if unorthodox ideas, in the present volume. And this is probably its most important feature.

Jagiellonian University
Kraków

NOTES

[1] John Watkins, *Science and Scepticism*, Princeton University Press, 1984.
[2] *Ibid.*

JOSEPH AGASSI

THE THEORY AND PRACTICE OF CRITICAL RATIONALISM

0. INTRODUCTION

1. It is better to seek the import of the different theories of rationality for the life of reason than to seek the response to the skeptic critique of rationalism (as even Popper, Bartley and Watkins do).

2. There are two modern theories concerning societies living the life of reason: some updated version of the near anarchism of Hume and Smith and some updated Platonist technocracy.

3. The need to answer the skeptic sounds reasonable because the opposite of justifying one's views is the caprice which is crazy, yet the reason that makes a view the opposite of caprice is far from being proof or proof-surrogate.

4. The theory of rationality may be that of a choice of or the belief in a hypothesis or a view (Parmenides, Plato) and it may be the choice of a community or of its lifestyle or of its system of beliefs (Hegel, Buber, Polanyi) and it may be the choice of and the attempt to improve upon the best standard of rationality available (Popper and, following him, Jarvie and myself).

5. The proposal (of Jarvie and myself) to view rationality as improvable is hard to stomach as it conflicts with the central idea of all traditional western philosophy, the dichotomy between nature and convention. But this dichotomy has to go as it advocated neo-anarchism or technocracy and leaves no room for the democratic way of life as the life of reason.

6. Though before the Enlightenment the theories of rationality did not pertain to the choice of a social order, they were not as indifferent to social matters as Weber assumed.

7. Democracy conflicts with both neo-anarchism and technocracy. Critical rationalism, Popperian or any variant of it, is essential for democracy, and even for much less: for *glastnost* and *peristroika*. It is also inherently social and political so that its democracy is not a corollary to it but built into it. Yet with the possibility of global annihilation it too is utterly outdated.

J. Misiek (ed.), The Problem of Rationality in Science and its Philosophy, 1–18.

1. WHY ANSWER THE SKEPTIC CRITIQUE?

It is better to seek the import of the different theories of rationality for the life of reason than to seek the response to the skeptic critique of rationalism (as even Karl Popper, William Bartley and John Watkins do). Rationalism is the concern with and the advocacy of rational thinking, of rational conduct in general, and of the life of reason in the most general manner conceivable. Somehow, this fact was lost in the thicket of the historical discussion about rationality (including the significant contributions of Popper and of Bartley, on which more will be said below): most rationalist philosophers are concerned these days with the answer to the skeptical criticism, the proof that rational justification is impossible, as if this is all that stands between rationality and perfection, as if this is the only problem left.

Perhaps this is not the best policy. Possibly one might want to spy on the promised land even before; after all, the effort has not succeeded despite centuries of hard work, and even the latest contributions were questioned (including Popper's by Bartley and Bartley's by Watkins, not to mention my own ignored skeptical criticisms of all three). Why should we wait, possibly for a few generations more, for the study of rationality that is not very germane to skepticism? And if the chosen land is not perfect, then possibly it is an error to invest all effort in the single, highly theoretical task, rather than seek some practical task. Indeed, many scientists, including Max Born, told me, rather spitefully, I am afraid, one is better advised not to waste time on the philosophical discussion of rationality and do research – to turn to philosophy only when one's powers are gone. Are they right? This matter should perhaps be put on the philosophical agenda. Or not, if philosophers can help solve practical problems of rationality here-and-now, assisting researchers in their work, promoting world peace, etc. Born's suggestion may apply to philosophers who respond to the skeptical critique, not to philosophers whose studies may help solve practical problems rationally.

There is an answer to this challenge: possibly the answer to the skeptic critique will be the best way to help solve practical problems rationally. Were this meant seriously, it would clearly imply that there was almost no progress in the applied field of the theory of rationality, since the answer to the skeptic critique is still not settled. Such progress as there is, if any, is interesting: *progress, regardless of whether scientific or metaphysical, may be pinpointed, examined and enhanced so as to examine its practical value, if any.*

This strategy will help decide whether the philosophers of the rationalist persuasion who persevere in the traditional attempt to answer the skeptics act rationally. Is it not wiser to invest some of their time on the question, what is the life of reason? After all, this question, too, has a tradition ever since Socrates was reported to have posed it in Plato's *Apology*. Moreover, the discussion of this question, what is the life of reason, just may be a better tool for the quest for the answer the skeptics than a direct attack on the skeptical argument.

2. WHAT CONSTITUTES A SOCIETY LIVING THE LIFE OF REASON?

Tradition presents a simple view of the individual's life of reason: it is the life of the individual devoted to study – perhaps to study combined with worship; in Jewish tradition, derived from Plato, no doubt, it is the life of worship where study is a cardinal item of worship. Spinoza integrated it into the modern tradition. In one sense it is unanimously endorsed and in another unanimously rejected. As the claim that rationality favors study, it is universally accepted. As the claim that there is no rationality except in study, it is generally rejected as much too narrow. Respect for learning is a fixture of all theories of rationality; each ancient school, the faithful, the Cynics, the Stoics, the Skeptics or the Epicureans, they all supplemented it with some proposals regarding the good life. Max Weber has noted that except for some of the religionists, they all made proposals which are unacceptable for the same reason: they speak of individual salvation, not of social progress: they are as oblivious to matters social as possible. This is only approximately true, as we shall notice, but let us endorse it for a while.

What then is the society living the life of reason? This question was asked by Plato and St. Augustine and it was rejected by the Judeo-Christian tradition as peripheral to the matter of the salvation of the individual soul. This question is central in the modern world, in whose tradition there are two popular (sets of) recommendations/answers to the task/question at hand, both rooted in ancient Greek tradition. The one is (Sophist) anarchism, comprehensive or tempered – the total or almost total independence of the individual, hopefully coupled with a attempts to advance one's material interests through a personal submission to science as best understood by the individual. The second is (Platonic) technocracy – the total submission to the expert scientist specializing in any matter that happens to be under consideration. (There are also the irrationalist answers to the question, what is the life which is worth living; the study of the life of reason can include them only after they are translated to rationalism; see below.)

The common form of anarchism is tempered: it allows that some minimal policing or another is advisable, so that the state is the night watchman of society, so-called. Its starting point is political: we are born free and we voluntarily relinquish some freedom – as little of it as is utterly necessary – for life within society. As I have argued in my *Technology* all anarchism is tacitly further tempered these days to one extent or another, by tacitly endorsing some compulsory education, national defense and the constraints on immigration, not to mention the coin of the realm. (It is no accident that Marx and Engels refused to discuss the question, how will the state wither away and into what will it be transformed: this discussion, Engels said, is utopian; yet he had a blind faith in the process of the waning of the state without discussing its mechanism; and he called his blind faith scientific. Since anarchists usually do not discuss these matters, it is only fair to ignore their doctrines here.

Technocracy is usually considered a variant of modern democracy, an enlightened variant, since it champions scientific technology as central to the modern way of life. I have argued in my *Technology* that this is an illusion for two strong reasons. First, the delegation of all important decisions to technicians abrogates all personal responsibility and turns all politics into expert public administration. Second, the coordination between technicians is a super-technique, whose practitioner is traditionally known as the philosopher-king. Popper has stressed the fact that Plato's philosopher-king is a scientist and argued that since after Einstein science can no longer be deemed free of error, in principle democratic leadership is to be preferred over philosopher-kings. This argument, it should be noted, is skeptical, not based on a reply to the skeptic. It is therefore regrettable that Popper says his philosophy is a reply to the skeptic.

3. JUSTIFICATION PHILOSOPHICAL AND COMMONSENSICAL

The reason the need to answer the skeptic sounds so reasonable is that the opposite of justifying one's views is the caprice which is crazy, yet the reason that makes a view the opposite of caprice is far from being proof or proof-surrogate. Traditional rationalism, the majority party within rationalism – were labeled by Sextus dogmatists (i.e., opinionated, not in the modern sense in which the word means, being attached to one's opinion to the extent of being unable to relinquish it come what may); Popper calls them comprehensive rationalists (i.e., requiring proof for every move, including proof of the validity of the proof procedure itself); Bartley calls them justificationists as they view rationality as comprehensive justification. The asset of the demand for rational justification of views is obvious: it is at least in intent an advocacy of life in the light of rationally justified ideas, allegedly the life of science – hopefully within democracy. Yet this rests on an erroneous view of science and of democracy, as Watkins has argued: it disallows disagreement – permitting it to appear only temporarily and marginally or on the forefront. (See my paper on the national consensus in the Watkins *Festschrift.*)

The strongest argument in favor of comprehensive rationalism is simple and powerful: one is not rational who holds views arbitrarily and/or behaves erratically and/or arbitrarily; such arbitrariness or waywardness may be called crazy; the sane, then, can provide some reasons for any choice of views, of conduct, and of a way of life; whatever reason is thus provided will constitute a justification or a rational defense of that choice. This is the central argument of all justificationists.

This argument is valid: the sane can give reasons and hence justify. Bertrand Russell stated it clearly enough. He declared the task of answering the skeptics imperative, as they contended that no justification is ever available. Russell also viewed anyone who hopes to show that empirical evidence plays some role in rationality – in rational thought, action, or way of life – as an advocate of a principle of induction, a principle of founding rationality on evidence. He therefore proclaimed Popper an advocate of induction; after Popper corrected him on this, he proclaimed Popper a defeatist (in his preface to the reprint of Nicod's book on induction).

To repeat, the valid and forceful claim at hand is that the sane can explain their choices of views/actions/ways-of-life; it is a strong argument in favor of justificationism. As such it is very general. The skeptic can only answer it, if at all, in a very general manner – by providing some general forms of counter-examples to it. Then the counter-examples may serve as touchstones to distinguish the justificationist and the non-justificationist forms of rationality. The counter-examples, too, are extremely simple: sane people may act rationally on unexplained convictions; they may be in situations which clearly invite action, even urgently, yet with no known line of action except ones known to be objectionable under the circumstances. It is then rational to try some new mode of action with no justification for it, other than that in the circumstances clearly any action is better than none; this claim is justification enough in the circumstances.

There is a clear choice here. There are very poor reasons, such as, we must act as best we know how. These may be considered justifications or not: on the demand that all justification should be scientific they are not. If not, then one may be not justified yet sane, since there are sane individuals not cognizant of science; alternatively, justification is the only alternative to craziness because it is so easy to provide a justification that one should always be able to pass the (psychiatric) test of it. Traditionally justificationism is the view that only science can justify, that most justifications offered by ordinary sane individuals who are ignorant of science are invalid.

We are here torn between two standards of rationality, one too high (all justification must constitute scientific proof) and one too low (the rational must pass a simple psychiatric test); confusing them is the base of the argument that made Russell take it for granted that we all justify except for the crazy, so that sanity equals justification. Justificationism recommends a standard of rationality to live by; psychiatric tests of sanity are based on the view of what practice is deemed rational by the community in which the test is applied; Russell tended to fuse the two! Popper accepted Russell's challenge to show that rationality is sane, and he then declared that he had answered the skeptic. But the skeptical critique is of the recommended standard, not the standard ascribed to a given community. And as Popper observes, the recommended standard should be discarded anyhow. Nevertheless, Popper agrees that, to begin, with the recommendation of any standard is open to the skeptical critique, and he therefore attempted to answer that critique. If he does, than he has justified the views he holds, if not, then his acceptance of Russell's challenge is the admission that he recommends insanity. Some justificationists, David Armstrong for one, uses this argument explicitly and emphatically; he considers me, for one, an advocate of insanity. David Stove considers Popper an irrationalist on this very ground. Yet the ordinary justification that ordinarily exempts people from the charge that they are insane is definitely less than the justification which justificationists ascribe to science.

The fact that commonsensical justification is not comprehensive and thus much inferior to the one required by justificationists, is a source of endless confusion (and of the inability to comprehend the fact, stressed by Sextus and by exasperated Pierre Bayle, that skepticism is no critique of common justification). One might have hoped that the philosophers of ordinary language and/or commonsense, would have cleared the confusion. They did not, as explained at some length in the paper on the philosophy of commonsense by myself and John Wettersten. This paper was not answered; perhaps uncritical rationalism is these days a mere ritual.

Justification is required in courts of law, where the level of strictness of proof is decided by convention (as discussed in my paper on the conventions of knowledge in Talmudic law). The legal question of justification is different from that of scientific proof and is thus not open to the skeptical argument.

Suppose that the justificationist philosopher will demand that rational justification should be the justification of responsibility and then identify the principle of induction with the legal principle of responsibility. This attempt to identify legal and scientific justification is outlandish. Admittedly, it should lead the philosophers in search of a principle of induction to law-courts. They will find out fast enough, then, that the law is different in times and places and is nowhere above criticism. The search for the universal rule of legal responsibility which is above criticism will then take them away from the law-courts and back to the search for the true, immutable principle of induction which justifies science as above criticism and which is thus other-worldly. The search for worldly standards of criticism will be empirical and show that different societies have different standards of rationality, and that in the same society there exist different standards of proof: capital punishment is allegedly based on evidence accepted in accord with the most stringent standards of proof and the police should prove grounds for suspicion of guilt by much more lenient standards and the law specifies still other standards – for journalists, legislators, diagnosticians and others. Some philosophers say, rational is whatever is deemed rational. If so, then the crazy is rational whenever they are listened to. The diversity of standards is much too wide and invites a universal one.

4. THE PROBLEM OF RATIONALITY SOCIALLY RECAST

The theory of rationality may be that of a choice of or the belief in a hypothesis or a view (Parmenides, Plato) and it may be the choice of a community or of its lifestyle or of its system of beliefs (Hegel, Buber, Polanyi) and it may be the choice of and the attempt to improve upon the best standard of rationality available (Popper and, following him, Jarvie and myself). It is quite aggravating that the problem of rationality can be presented both most abstractly and most concretely, as these two approaches are often confused. Abstractly, what thought/action/way of life is rational? Concretely, as there are different canons of rational thought/action/way of life, how are we to choose between them? The abstract approach is traditional ever since Plato presented comprehensive rationalism: rational thought is scientific proof, rational action is based on rational thought, and the rational way of life is the life of rational thought and action.

The concrete approach to the problem of rationality developed in diverse attempts to eschew Plato's prescription, perhaps chief among them is due to Hegel. It is an error to suggest that Hegel had a new critique of Plato's rationalism. His critique was of his immediate rationalist predecessors who had added to Plato's views some practical suggestions: it was the liberal social component of their suggestion which Hegel criticized. His criticism of justificationism as such does not go even to the length of the skeptic critique (see the end of the Introduction to his *The Phenomenology of the Spirit*, passages on different kinds of science and on science as a phenomenon). He rejected the idea that the rational way of life is in the light of science and open to all; he advocated traditional society while rejecting the utter stagnation advocated by traditionalists. He dismissed all ideas as insignificant and discussed action, especially political action, especially big, impressive political action, including large-scale military fireworks, which he advocated as an antidote to stagnation, as carriage for progress. He claimed that hardly anyone has the option and exercise the choice to move from their native societies into other societies, not to mention create new societies. Yet he did permit some geniuses here and there to alter tradition – for the better, of course. And geniuses are moved by ideas, he said, but not by thinking them out; rather they are moved by irrational passions, by the [infantile] ambition to stand out. Choice, according to Hegel, it is made by impulse, by gut-feelings.

This idea was mulled over many times and there were many variants of Hegel's action philosophy, as both Russell and Popper called it, some coarse, others refined, and all share the proposal that the choice is made irrationally and yet we are all the better off for this.

The most urbane variant of action philosophy is due to Martin Buber and it was endorsed by Michael Polanyi (the latter did not refer to the former as far as I know, but he used the expression "I and thou" which is his fingerprint, of course): we can choose only which community to join, and we better choose by attending to the cultural element it offers. Buber had in mind a religious community and a society that can offer a rich and satisfying communal way of life (he followed Gustav Landauer and was a communalist-anarchist). Polanyi was eager to join both a religious community and a scientific community. Doctrines clash and so do ideologies; lifestyles do not; presenting both religion and science as lifestyles he overcame the traditional conflict between science and religion in one stroke. A variant of Polanyi's philosophy is these days advocated very successfully by Thomas S. Kuhn and the debate raged around the allegation that Kuhn is an irrationalist since he thinks that both joining the community of science and endorsing the revolutions within science are not rationally justified. It is obvious that Kuhn is an irrationalist who follows Polanyi in advocating the view that scientific justification is neither possible nor necessary. Unlike Popper and his entourage, Kuhn has never contemplated a critical, non-justificationist rationalism. The idea that the problem of rationality may be better posed concretely as the choice between life-styles is still the precinct and purveyance of the irrationalists.

Yet the concrete choice of one community or of one standard of rationality among the extant ones need not be as irrationalist as initially intended. This was emphasized by Ian Jarvie and myself in a series of papers we have co-authored, reissued in the volume *Rationality: The Critical View* which we have co-edited. It is very hard for some philosophers to absorb this, as indicated by most of the many answers which our papers were fortunate to receive. And it is an easily understandable difficulty. Admittedly, the standard of rationality is a standard of propriety of sorts, and we are well aware of the fact that different cultures offer different standards of propriety. Yet endorsing the different standards simultaneously seems relativist and so irrationalist, and endorsing rationalism is admitting that rationality is a peculiar, universal standard. The recognition of the obvious fact that there are different standards of propriety and its import for our choice of a rational way of life is the very core of western philosophy: the fire burns the same way everywhere but the law of the land differs from place to place, said old Protagoras. This observation signified as it instituted a search for a better standard than that justified by local tradition: the search for rationality was the search for a universal standard of rationality akin to the universal laws of nature. Hence there can be no diversity of criteria of rationality; the claim Jarvie and I make is thus absurd and Plato's solution is the only possible rationalist one. This argument, presented against Jarvie and myself, seems impeccable. It should force us to give up our views or to reject one of its presuppositions. Moreover, if our answer is not to be an attempt to wriggle out, it must include the rejection of some significant premise. It does.

5. REPLACING THE CLASSICAL DICHOTOMY WITH A TRICHOTOMY

The proposal of Jarvie and myself is hard to stomach as it conflicts with the central idea of all traditional western philosophy, the dichotomy between nature and convention, the division of all truths into truths by nature and truths by convention. This dichotomy is ubiquitous: it appears in discussions of science, of morality and of politics. In science it limits our choice to utter arbitrariness or to utter dogmatism, leaving no room for the freedom of rational dissent. In ethics it limits our choice to two options, the utter disregard of conventions and utter conformism, neither of which leaves any room for moral reform. In politics it limits our choice to two options, neo-anarchism and technocracy, neither of which leaves any room for the democratic way of life as the life of reason.

In my *Towards a Rational Philosophical Anthropology* I have presented this dichotomy as grandiose but obsolete, as the source of western philosophy or its major distinctive characteristic but as one western philosophy must now relinquish, as leading to justificationism and to reductionism which have made great contributions in the past, and now must go and leave room for better metaphysics and methodology.

The rejection of all truths by convention is called naturalism. The claim that there are significant truths by convention has repeatedly been declared the unfailing source of cynicism (in the modern sense of the word, signifying disrespect for the truth), and perhaps even the view that leads one by inexorable logic to cynicism and to nihilism. This seems to be a view which Plato has advocated and it has been repeated in a magnificent manner by Leo Strauss in his, *magnum opus, Natural Rights and History*. It seems clear that when Russell declared the search for the solution to the problem of induction, he was troubled by the cynicism of the conventionalists (those who declare that some important truths are truths by convention); yet he was equally troubled by naturalism as justifying dogmatism. He found the dichotomy a dilemma and he confessed he could not break away from it.

Sir Karl Popper has honored me in his *Replies to My Critics;* in the passage devoted to my criticism, he declared that he does not endorse the dichotomy in question. I am proud that I gave him the occasion to state this in a clear and unequivocal manner. In his grandiose *The Open Society and Its Enemies* of 1945 he attacks naturalism and concludes that conventionalism is correct: he declares himself a moral conventionalist. In his *Logik der Forschung* of 1935 he rejects both naturalism and conventionalism and admits the existence of conventional elements in science. This admission plus the dichotomy should make Popper a conventionalist. A number of writers declared him a conventionalist – not unreasonable in view of the absence of a clear rejection of the dichotomy in that book. *The Open Society and Its Enemies* proposes that social institutions are hypotheses; hypotheses they are true or false, yet the dichotomy designates institutions truths by convention. A decade later Popper came out with his trailblazing "Three Views Concerning Human Knowledge" (1956; reprinted in his *Conjectures and Refutations*, 1961). The dichotomy is now clearly replaced by a trichotomy, yet the claim that the Presocratic dichotomy is erroneous is stated neither there nor in its companion "Back to the Presocratics". The nearest Popper came to an explicit rejection of the dichotomy before his *Replies* is his "On the Sources of Knowledge and Ignorance" (p. 18 of his *Conjectures and Refutations*) where he rejects the theory of truth as manifest, the idea that nature is truthful and error is due to human convention.

Nor is the matter so straight forward: the trichotomy is not as straight forward as it looks: it is easy to envisage truth by nature, representing the reality behind the veil of the appearances, and it is easier to comprehend truth by convention, the truth of mere working definitions; but what are the levels of reality that correspond to the levels of approximations to the truth? It is clear enough that there are levels of approximation in science; but do the theories on each level represent real layers of facts? No: what is merely a matter of great precision in one environment is a colossal difference in another. For example, there is a tendency to view matter as continuous on a superficial level, as atomic with stable atoms on another and the atoms disintegrating very seldom so that one needs special sensitive instruments like Geiger counters to detect it; not so: as the sun is a nuclear furnace, in it things are totally different, as they are to a smaller extent in atmospheric nuclear explosion. This is a tremendous difficulty and it explains the absence of a discussion of nature and convention in "Back to the Presocratics" as well as the discontent with Popper's theory of verisimilitude that goes well beyond any critical debate of any technicality involved in the precise definition of verisimilitude. All this, however, is neither here nor there: what I wish to report here is that one way or another, the ancient Presocratic dichotomy is now a thing of the past.

In my note for Paul Arthur Schilpp's *The Philosophy of Karl Popper*, incidentally, to which Popper was replying to say that he does not endorse the classical dichotomy, I suggested that Popper's philosophy is better characterized as modified conventionalism, since for him the accepted truths, the instituted truths, the traditional truths, are not truths by nature, yet they are not arbitrary either. I failed to connect. I should have said then, I suppose, that my discussion concerned not rational thought or action but the society in which the rational way of life is practiced. What kind of society practicing the rational way of life should a rationalist philosopher advocate and how is that advocacy linked to a theory of rationality?

6. THE SOCIAL COMPONENT OF TRADITIONAL PHILOSOPHY

Before the Enlightenment, to repeat, theories of rationality did not pertain to the choice of a social order but to individual salvation. Nevertheless, they were not as indifferent to social matters as Max Weber has assumed. He took for granted comprehensive rationalism, or justificationism, and was thus limited to the choice between anarchism and technocracy. Incidentally, the following compromise between technocracy and anarchism is possible: individual tastes are permissible even if the technocrat should recommend to change them. The presupposition here is, of course, that the public can afford to defer to individual tastes in the matter under consideration. This compromise or something of that order has been recently offered by Paul K. Feyerabend – in his reply to my review of his first crazy book that appears in his second and crazier book. This compromise, to repeat, is on matters of public luxury; it is typical of those living in the luxury obtainable in exchange for their disregard of the grave problems which humanity is facing these days. Since the compromise is marginal we should now ask, what other social orders and other views of rationality are there?

One may object: there is no obligation for a theory of rationality to prescribe a social order and a way of life: the number of ideas concerning rationality is limited only by the imagination, whereas the number of social orders is limited by historical fact. This, we remember, is why the irrationalists prefer to discuss the choice of ways of life of given social orders: this choice concerns attitudes to already existing social orders: it is both more manageable and more likely to be reactionary. This is a major philosophical constraint on the practical discussion of rationality and the task is to accept the reactionary attitude while diffusing its reactionary character. The pioneering effort in this direction is outlined in Popper's "Towards a Rational Theory of Tradition" (reprinted in his *Conjectures and Refutations*). How does one choose from among the given ways of life without being an irrationalist?

The story is that the skeptics had nothing to offer in this matter. Yet the radicals had nothing more to offer, perhaps. The radical recommends the social order allegedly based on science. This is the modern world – allegedly or in truth – and it did not exist before World War II, or, at the most generous estimate, before the industrial revolution. How did radical and skeptic differ before that? The answer seems to be, inasmuch as there was a difference, it was in the hope, that the radical had and the skeptic lacked, to see the immanent rise of scientific society (Sir Francis Bacon). What was the skeptic offer?

Perhaps the skeptic had nothing to offer. Sextus and Bayle answered the claim that skepticism is un-commonsense and thus absurd, by the claim that the skeptic is none the worse than the dogmatists, since no theory offers help here, except that whereas the dogmatist is tormented by doubt the skeptic savors it and the peace of mind which it provides. After all, Stoicism, too, was a doctrine full of personal advice offered in disregard for the social order. And, indeed, what social order did any philosophers, the dogmatist, the Stoics, the skeptics or the religious devout, what social order could they offer?

This is a rhetoric question, and as such most unfair. The early dogmatists were turned on by their courage: dismissing all truths by convention they were opening the road to the true unity of the human race. One may speculate about the motive that made Alexander the Great make his generals marry local princesses: after all, generals and princesses have truth by convention to thank, as their very distinction depends on it. But it is not easy to overlook the fact that philosophers did attempt to live in the midst of their own societies in disregard for the truths by convention of their time and place, that some of them were exceptionally brave, that the accusation that philosophers were demoralizers was not baseless.

The word which W.W. Bartley chose to designate the dogmatists by, "jusitificationists", is the precise translation of the ancient Hebrew word "Sadduceans". When I teach in Israel and use this word, my students cannot take it; they use different vowels to differentiate the ancient Sadducean theory from the modern one. The reason is plain: they learn philosophy from teachers who are justificationists, and so they tend to view justificationism with favor; even when they accept some criticism of justificationism, Popper's or Bartley, even when as a result they are willing to desist from justificationist attitudes, they are left with a favorable flavor for it. They learn about the old Sadduceans from their Pharisean teachers, and the Phariseans were never renown for a decent presentation of their opponents' views: they never admitted that the (inter-testament) Sadducean tradition recommended the endorsement from Jewish tradition of only what could be justified in the light of reason; rather, they presented the Sadduceans as people ashamed of their heritage, as those who relinquished the word of God in preference for Greek conventions, such as sports. It is not that in Israel there is no deep disagreement about sports. But the radicals who founded the new Israeli society have established physical culture as a part of the modern reality there and the link of this with ancient debates, the fact that the modern debate is in a sense a repeat performance, is for Israeli students unknown or too disturbing or both.

The Sadducean way of life is not quite practicable prior to the industrial revolution – a point repeatedly and forcefully made by Ernest Gellner. No philosophy, he observes, could offer the social component required for the enlightened choice of a social framework as a frame for a reasonable way of life. The reasonable choice for the individual always was – as Martin Buber has stressed in all of his writings – the choice of ways for integrating in a given society, or in the best subculture it offers, or something to that effect. At times the choice was heartbreaking, as in the choice between staying Jewish or joining the promise of Enlightenment, or even between staying in one's tribe or moving to the modern world of plenty. Clearly the most exciting is the most basic choice: between truth by nature and truth by convention as offered by the Presocratics. Even in the modern world, Jews were more attracted to the promise of the naturalists than to the plenty that the conventional Christian world could or did offer them.

This is a point that I found only in one place stated, and very forcefully, yet it seems with little notice from the learned world: in his historical writings Sir Lesley Stephen, a critical rationalist, I think, has noticed that comprehensive rationalism, the justificationist optimism of Spinoza and Leibniz, was a great polarizer: a doctrine very difficult to practice properly and very handy to excuse, to justify, helplessness and inaction and folly of all sorts.

7. RATIONALISM IS A DEMOCRATIC SOCIAL PHILOSOPHY *AB INITIO*

Democracy conflicts with both neo-anarchism and technocracy. Critical rationalism, Popperian or any variant of it, is essential for democracy, and even for much less: for *glastnost* and *peristroika*. It is also inherently social and political so that its democracy is not a corollary to it but built into it. Yet with the possibility of global annihilation it too is utterly outdated. Today justificationism is triumphant, so that to live by it is no longer the titanic challenge it used to be. Admittedly, the paradigm of rational justification of a doctrine was the justification of Newtonian mechanics, and this has been overturned. This calls for some fence-mending. Perhaps there is a need for a justification that will allow a revolution now and then to keep science on its toes, perhaps the scientific community can demand loyalty no different from other societies, as Michael Polanyi has observed. One way or another, it seems foolish to give up the venture just when science has come of age and has been able to offer a rational way of life. If we managed without having quite won the battle with skepticism for so long, and if we have so magnificently succeeded till now, surely we can try a few decades more.

What is wrong in the analysis presented here is not so much that the skeptical critique of justificationism is unanswerable. Everyone knows that it is unanswerable, but few are as impressed by it as the critical rationalists. All logical arguments are forceful at the cost that all logic has to pay: the slightest alteration in the doctrine logically refuted may totally immunize it to the refutation.

What is wrong in the analysis presented here is the *ad hoc* way in which the social component has been introduced into the discussion. The problem, the choice, the concern, all were of the individual: society had no role to play in it except that of a problem for the individual. Perhaps even the individualism that is so characteristic of western society, culture, philosophy, is but a by-product of the appeal of the Greek ethos to the individual – simply because it was an appeal to be critical of conventional wisdom. (The logic of this situation is obvious in Hebrew tradition: prophets who attacked convention and had nothing stronger to fall on she or he tended to appeal to individual conscience. Be it as it may, since the problem and discussion center on the individual, the social component of the individual's choice entered it as an after-thought. This will not do.)

The situation as described in the previous paragraph explains the odd fact, alluded to in the beginning of this essay, that democracy is not easily supported by any traditional theory of rationality, that critical rationalism is its foundation well after the fact. Democracy cannot be introduced into philosophy after the fact except if it is an appealing and strong existing option.

Where does society enter the discussion "naturally"? At the very start: at the problem-setting. The fire burns the same way everywhere but societies differ. What if it? Why did this trouble the ancient Greek sophists so much? Because they discovered options they had not known exist. Herodotos narrates that the different social options were not noticed because all customs other than one's own were "naturally" considered inferior; that there was a blind spot that was to be rid of before this was discovered: as he explains by narrating a myth: the magi had to be expelled before rationality could be instituted. Now all this may have been revolutionary in Antiquity and even in some not so distant past, but it is today neither very new nor very exciting. However difficult it is for an individual to break ties with tradition, it can be done, and at known cost, and while raising certain social and political and educational problems, not philosophical ones. No matter which of the different extant attitudes to custom is endorsed, it is philosophically rather unproblematic. Why? Because we have learned to circumvent it: out of so much frustration, perhaps, we have found that a social problem is best attacked by new technology rather than by forging and/or implementing new attitudes.

In my book *Technology*, I tried to show that this way bottlenecks are not opened but postponed, and at high interest rates. The long and the short of it is that now all traditional societies, primitive, advanced, and in between, are helpless in the face of the coming cataclysm that the ecological and the peace movement are discussing at such a great and colorful detail. If the future is anything like the one predicted, we are doomed. If the ecosystem will not be destroyed by the chemical poison the industries of the whole world pour into it or by the enormous waste of fossil fuel, then it will be destroyed by the population explosion. Alternatively the population explosion will lead to reckless wars, perhaps with the use of nuclear weapons; and even if the population growth in the poor parts of the world can be contained, there is no reason to assume that the poor nations will be willing to tolerate the *status quo*. In brief, if ever a new plan – as new as our boldest imagination can offer – is wanted, it is now. There is no way back and slow progress is as dangerous as the *status quo* or more. And what are the philosophers doing about it? What they have been doing all the time and less.

What can be done? How is the proposal to do anything related to any theory of rationality? I do not know. If there is any rhyme or reason in current global politics, it is its being so very marginal: almost all global politics is performed in the service of local interests: politicians pay heed to global politics only through the prism of national and narrower interests. Few politicians ever have the resourcefulness and vision to develop long range plans, anyway. Progress nevertheless took place, of course: politicians from time to time did recognize the requirement to come up with long range plans and engage increasingly wider outlooks. Otherwise it would not be possible to develop the modern nation-states out of small local interests. Once a politician saw the advantage of a wider outlook, of planning for larger systems, the ensuing success forced competitors to follow suit.

This is no longer the case. If the whole world will not mobilize in order to put global issues high on their agendas, then nothing will ensue: the global problems of the survival of humanity can only be solved globally, and local solutions will not begin to touch them. If anything, if some but not all parties are reluctant to use resources for fear of harming the globe as a whole, then those who do not share the reluctance benefit from it. How then can the world leadership on all of its variety agree that certain moves are essential for global survival? How can we ever hope to alter the habits of millennia and invite more rationality in global politics before it is too late? Is it rational to exhibit a reluctance to use certain resources for fear of extinction? Can claims for the existence of a causal relation between the overuse of fossil fuel and extinction be tested? How much testing is required? Of which kind? Is that not the traditional problem of induction run wild?

Assuming that the present problematic situation is a case of an inability to apply justificationism, or, more specifically, the principle of induction, one admits that inductivism is now bankrupt, regardless of what is claimed for it concerning the past. The question remains, what is the situation from a critical rationalist viewpoint?

There is a clear sense of crisis in the air. Whether the future of humanity is gloomy or not, it clearly is severely threatened. If survival is the topmost item on every practical agenda, then the survival of humanity should be the topmost item of every agenda, private, philosophical, national, global. Does any philosophy address this matter? Not that I know of. Back to the drawing board, and fast: the time may be too short, or we may just barely come up with a solution in the nick of time. For my part, I can only pose the problem. What I have ventured to illustrate is that the concern for a rational social order was never far from the deliberations on rationality, that these days we have to be more explicit about it for intrinsic reasons and make it global and paramount for extrinsic ones.

The problem of rationality has radically transformed: it reads, how can we save ourselves from destruction? How is global security to be achieved?

Tel-Aviv University and
York University, Toronto

REFERENCES

Agassi, J., 1974, "Conventions of Knowledge in Talmudic Law", in B. Jackson, editor, *Studies in Jewish Legal History, in Honor of David Daube, J. Jewish Studies*, 25, 16–34.

Agassi, J., 1977, *Towards a Rational Philosophical Anthropology*, Kluwer, Dordrecht.

Agassi, J., 1981, *Science and Society: Studies in the Sociology of Science, Boston Studies*, 65.

Agassi, J., 1985, *Technology: Philosophical and Social Aspects*, Kluwer, Dordrecht.

Agassi, J., 1988, *The Gentle Art of Philosophical Polemics*, Open Court, LaSalle IL.

Agassi, J., 1989, "The Logic of Consensus and of Extremes", in F. D'Agostino and I.C. Jarvie, editors, *Freedom and Rationality: Essays in Honour of John Watkins. Boston Studies*, 117, 3–21.

Agassi, J., 1990, "Global Responsibility", *J. Applied Phil.*, 7, 217–221.

Agassi, J., 1991, "Deconstructing Post-Modernism: Gellner and Crocodile Dundee", in John A. Hall and I.C. Jarvie, editors, *Transition to Modernity: Essays on Power, Wealth and Belief*, Cambridge UP, Cambridge, 213–30.

Agassi, J., 1991, *Review of World Commission on Environment and Development*, "The Brundtland Report", *Intl. Rev. Soc., Monographic Series*, 3, 213–226.

Agassi, J. and I.C. Jarvie, 1987, *Rationality: The Critical View*, Kluwer, Dordrecht.

Agassi, J. and J.R. Wettersten, 1987, "The Philosophy of Commonsense", *Philosophia*, 17, 421–437.

Bartley, W.W., 1984, *The Retreat to Commitment*, 2nd edition, Open Court, LaSalle IL.

Buber, Martin, 1957, *Pointing the Way*, Routledge, London.

Buber, Martin, 1960, *Paths in Utopia*, Beacon Press, Boston.

Feyerabend, Paul K., 1978, *Science in a Free Society*, NLB, London.

Nicod, Jean, 1950, *Foundations of Geometry and Induction*, Routledge, London.

Polanyi, Michael, 1964, *Science, Faith and Society*, Chicago University Press, Chicago.

Popper, Karl R., 1959, *The Logic of Scientific Discovery*, London, Hutchinson.

Popper, Karl R., 1959, *Conjectures and Refutations*, London, Routledge.
Schilpp, P.A., editor, *The Philosophy of Karl Popper*, Open Court, LaSalle IL.
Stephen, Sir Lesley, 1904, *History of English Thought in the Eighteenth Century*, Duckworth, London.
Strauss, Leo, 1965, *Natural Rights and History*, Chicago University Press, Chicago.
Weber, Max, 1964, *Sociology of Religion*, Beacon Press, Boston.

FRED D'AGOSTINO

A DEONTOLOGICAL APPROACH TO THE RATIONALITY OF SCIENCE

How are we to characterize the rationality of science? It is usual to suggest that science, like many human enterprises, has an aim and that the activities of its practitioners are rational to the extent that they promote the realization of this aim. Scientific methodology, on this account, involves determining how this aim is to be optimally pursued. Whatever distinguishes science from other human activities lies in the peculiar nature of the aim of this activity. For insofar as it is rational, its rationality is just the *instrumental* rationality that is familiar in other contexts – and consists, in particular, in the adaptation of the means adopted to the aim pursued. This view is very widespread.[1] I reject this view. Following up a hint by John Harsanyi, I aim to develop here an alternative to the instrumental account of the rationality of science which has for so long prevailed unchallenged.[2] Adapting some insights from contemporary political theory, I aim to develop a *deontological* account of the rationality of science.[3] I begin by considering some possible *attitudes* to the existence of disagreement – for it is, in my view, in response to the fact of disagreement that the need for a method of science in fact makes itself felt. I consider relativistic, dogmatic, and 'liberal' attitudes to disagreement, and I try to show that the liberal attitude is the *only rational attitude* to adopt in the face of disagreement. I then consider the *responses* to disagreement which are appropriate for those who have, rationally, adopted a liberal attitude to it.

I. THREE ATTITUDES TO DISAGREEMENT

What's involved in adopting a relativistic attitude to some disputed proposition? To say that Beth is a relativist with respect to some proposition *p* (which she affirms, let us suppose) – to say, in short, that she is a *p*-relativist – is to say, in effect, that Beth recognizes this possibility: viz. that Alf might with good reason deny that *p* without there thereby being any good reason for her also to deny that *p*.

The relativist in other words insists, contrary to the *universalist*, that what is reason-for-one to deny (or affirm) need not be reason-for-the-other to deny

J. Misiek (ed.), The Problem of Rationality in Science and its Philosophy, 19–43.

(or affirm).[4] What counts as a reason for denial (or for affirmation) is on this account relative (to the person who denies or affirms).

The universalist recognizes, with respect to some disputed proposition *p*, only two possibilities:

1. Beth with good reason affirms that *p*, and Alf (*ipso facto*) without good reason denies that *p*, or
2. Alf with good reason denies that *p*, and Beth (*ipso facto*) without good reason asserts that *p*.

The relativist recognizes, in addition to these two possibilities, the possibility that

3. Beth with good-reason-for-her affirms that *p*, and Alf with good-reason-for-him denies that *p*.

Now it cannot honestly be asserted that a relativistic attitude is never an appropriate attitude to discovered disagreement. Questions about so-called 'matters of taste' are probably best approached from a relativistic point of view.[5]

Of course, an attitude of *aestheticism* is not hereby validated. It would be wrong, surely, to claim, as does the aesthete, that all disputed propositions, like the 'matters of taste' illustrated at note 5 above, are immune to rational argumentation.[6] Imagine that Alf and Beth disagree about the best way to get back to their car. Alf wants to take the right-hand path, Beth the left; both want to return to the car as quickly as possible. Now if the claim that the left-hand path is considerably shorter than the right-hand path is a reason-for-Beth to prefer the left to the right (as this claim would be were it true), then, surely, that very claim is also a reason-for-Alf to prefer the left to the right; it is, in short, a reason-for-all or, perhaps, a reason *simpliciter*. And these considerations, as simple as they are, show, surely, that an attitude of aestheticism cannot rationally be sustained. (This fact will be of some significance in what follows.)

What's involved in adopting a dogmatic attitude to some, let us suppose disputed, proposition? To say that Alf is a dogmatist with respect to some proposition *p* (which he affirms, let us suppose) – to say, in short, that he is a *p*-dogmatist – is to say, in effect, that Alf is disposed to believe, of any other proposition *q* which is logically incompatible with *p* (i.e. whose truth would imply the falsity of *p*), that this proposition cannot possibly be true, if on no other grounds then at least in virtue of its incompatibility with *p*. Furthermore, Alf asserts that he has, with respect to such a proposition, an *incontestable* first-person authority to determine solely on the basis of his own experience whether this proposition should be affirmed or, instead, denied, and on this basis claims the right to reject those considerations presented by others as reasons for denying that which he affirms. The dogmatist in other words insists, contrary to the *fallibilist*, that it sometimes suffices, to warrant an attitude of affirmation or denial, simply to consult one's own experience and that considerations apparently at odds with the attitude adopted on this basis reveal their infirmity in the very fact of their incompatibility with such experience. And this means that *nothing* that Beth can say can possibly provide Alf with a reason, which he will accept as such, for believing that *p* is false.

Suppose, for instance, that Beth presents as grounds for abandoning *p* an argument of this general form.

(1) *q* is incompatible with *p*; i.e., one of *p* and *q* must be false.
(2) *q* is true.
(C) Therefore, *p* must be false.

If Alf is a dogmatist with respect to *p*, he will reject this line of reasoning as infirm (i.e. unsound) despite its cogency. He will say, in particular, that *q* could not possibly be true, in which case he need not reject *p*, even though he would have had to, assuming the truth of (1), if *q* had indeed been true. For Alf believes, being a dogmatist with respect to *p*, that any proposition which is incompatible with *p* cannot possibly be true, and that the proposition *q* mentioned by Beth as grounds for rejecting that *p shows its falsity* in the very fact of its incompatibility with *p*.[7] Alf reasons, in other words, as follows:

(1′) Any proposition which is incompatible with *p* is, necessarily, false.
(2′) *q* is incompatible with *p*.
(C′) Therefore, *q* is, necessarily, false.

Rejecting premise (2) of Beth's argument as false (necessarily), Alf continues, dogmatically, to believe that *p*. (If Beth's argument against *p* does indeed depend on a false premise then it establishes no rational grounds for rejection of *p*.)

Notice that this analysis is completely general: it applies to *any* proposition which is incompatible with *p* – with respect to each such proposition Alf will, if he is indeed a *p*-dogmatist, respond as he has with respect to *q*: rejecting each as false on the grounds, if no others are available, of its incompatibility with the proposition to which he is dogmatically committed.

Now, it cannot honestly be asserted that a dogmatic attitude is *never* an appropriate attitude to discovered disagreement. Questions about a person's own psychic states are perhaps best approached from a dogmatic point of view.[8]

Of course, an attitude of *authoritarianism* is hereby not validated. It would be wrong, surely, to claim, as does the authoritarian, that *all* disputed propositions are, like such first-person claims about psychic states as were illustrated at note 8 above, immune to rational argumentation in view of an individual's incontestable first-person authority. Recall the dispute between Alf and Beth about the best way to get back to their car. Alf wants to take the right-hand path, Beth the left; both want to return to the car as quickly as possible. Now Alf might claim to know with incontestable first-person authority that the right-hand path is shorter than the left, but his claim is, surely, much, much less plausible in a case of this kind than in cases where his own psychic states are, instead, at issue. *Authoritarianism cannot be sustained.* (This fact will be of some significance in what follows.)

What's involved in adopting a 'liberal' attitude to some, let us suppose disputed, proposition?[9] To say that Alf and Beth are 'liberals' with respect to some proposition *p* (which she affirms and he denies, let us suppose) – to say, in short, that they are *p*-liberals – is to say, in effect, that each recognizes this possibility: viz. that the other might have good reason for adopting to the disputed proposition the attitude that s/he has in fact adopted and that such a reason, if it exists, if a reason-for-all and not merely, as the relativist might have it, a reason-for-some.

Liberals are, in other words, fallibilistic universalists. Unlike the relativist, they embrace a universalistic understanding of reasoning, believing, in particular, that what is a reason-for-one to affirm (or deny) is *ipso facto* a reason-for-all (and therefore, in particular, a reason-for-them) to affirm (or deny). Unlike the dogmatist, they embrace a fallibilistic understanding of human experience, believing, in particular, that their own experience provides them with no incontestably authoritative basis for affirmation or denial, but must, instead, be tested against the experience of others.

So, where the dogmatist recognizes, with respect to some disputed proposition *p*, only one possibility:–

1. that I with good reason assert that *p*, and that my 'opponents' (*ipso facto*) without good reason deny that *p*,

while the relativist recognizes in addition two other possibilities:

2. that I without good reason assert that *p*, and my 'opponents' (*ipso facto*) with good reason deny that *p*, and
3. I with good reason-for-me assert that *p*, and my 'opponents' with good reason-for-them deny that *p*,

the liberal recognizes in addition to 1 only the possibility 2. (The liberal is, I repeat, a fallibilistic universalist.) If there is disagreement, then someone must be in error. If the disagreement can rationally be resolved, then it can be resolved only by discovering a *reason for convergence* (i.e. a reason for asserting or denying that *p* which is a reason *both* for Alf *and* for Beth – and for anyone else concerned with the question; which is, in other words, a reason *simpliciter*, rather than a reason-for-some-but-not-necessarily-for-others).

As I have already pointed out, it cannot honestly be asserted that relativistic or dogmatic attitudes are never appropriate attitudes to encountered disagreement. It can nevertheless reasonably be claimed, or so I will argue, that *only a liberal attitude to disagreement is, in general, a rational attitude*, and, therefore, that one cannot, except irrationally, adopt to discovered disagreement a relativistic or a dogmatic attitude.

To establish this conclusion, so important for all that follows, I need, first, to give a general characterization of *rationality*.

I will say, then, that an individual is rational in relation to a particular proposition *p* only if s/he does, when presented with good reasons for doing so, alter the attitude s/he has adopted with respect to *p*. (The basic idea is that the rational individual's attitude to some proposition is grounded in reasons and *varies as they do*.)

One is not rational in the attitude that s/he has adopted who affirms or denies irrespective of relevant reasons (i.e. who continues to affirm even when grounds for denial have been presented, who continues to deny even when grounds for affirmation have been presented). If someone cannot, no matter how good the reasons we offer, be induced, by an appeal to those reasons, to adopt a different attitude with respect to some particular proposition, then such a person is irrational in the attitude s/he has adopted to that proposition: s/he is rationally immovable with respect to the proposition in question (though not perhaps in any other way).[10]

Imagine now that Alf and Beth disagree about some proposition *p*, Alf affirming that *p* and Beth denying that *p*.

If Beth is a relativist with respect to *p*, then it cannot be guaranteed that Beth's attitude with respect to *p* will be a rational one in the sense of the preceding paragraphs. For Beth might claim, consistently with her denial of universalism, that there is, despite Alf's presentation of a good reason to affirm that *p*, no good reason-for-her to affirm that *p*, and thus might continue, in the face of a good reason not to do so, to deny that *p*. Her attitude to *p* might not, in other words, vary in accordance with reasons and might not, therefore, be a rational one.

Great care is needed in the formulation of this argument, especially in view of my admission that Beth might sometimes be right to adopt to the disputes she faces a relativistic attitude (as, e.g., in 'matters of taste'). I am not claiming – it would be wrong in view of this admission to do so – that Beth's attitude is *always* inappropriate, and that she *always* irrationally distinguishes reasons-for-Alf from reasons-for-her to deny that *p*. What I am claiming is that she is, whereas a liberal is not, vulnerable, unless aestheticism is accepted as true, to a particular kind of mistake – viz. that of thinking that the issue in dispute is, when it is not, one to which a relativistic attitude is indeed appropriate.

Let's proceed slowly here. Unless aestheticism is true, some issues are and some issues are not ones to which a relativistic attitude is indeed an appropriate one. But disputed issues do not come 'labelled' as ones to which a relativistic attitude is or is not an appropriate attitude. Beth is therefore vulnerable to error when she claims, in a world in which aestheticism is not true, that the dispute about the proposition *p* is in fact an issue with respect to which a relativistic attitude is the appropriate attitude: she might be right about this, but she might also be wrong. If she is wrong, then she wrongly adopts a relativistic attitude and as a consequence wrongly rejects as reasons-for-him-but-not-for-her Alf's grounds for affirming that *p*. (If the relativistic attitude is the wrong one, then Alf's reasons for affirmation are reasons which ought to persuade Beth as well to affirm; they are reasons, in other words, for altering the attitude to *p* which she has in fact adopted.) But Beth in this case fails the test for rationality: she does not vary her attitude in accordance with reasons to do so. She has failed in rationality precisely because she has (wrongly) adopted to the proposition in dispute a relativistic attitude.

Notice, in addition, that the 'liberal' is not subject to a mistake of this kind, and that liberalism is thus superior to relativism on this basis as an attitude to encountered disagreement. Since liberals are universalists, they will always treat reasons-for-others as if they were reasons-for-them-too and so will not mistakenly refuse to be swayed by such reasons for altering their attitudes to disputed propositions as are in fact available.

Now you might think that liberals, who are as I've said universalists, are vulnerable to a mistake which is, as it were, the 'mirror-image' of that to which the relativist is subject – and, therefore, that what is gained on the roundabouts is lost on the swings. (You might think, in other words, that liberals are just as vulnerable to mistakes, though to different kinds of mistakes, as relativists are.) Remember that I said that the relativist *will sometimes be right* to adopt to encountered disagreement a relativistic attitude; and, therefore, to distinguish reasons-for-one from reasons-for-the-other; and, therefore, to refuse to see in a reason-for-some-other for affirming that *p* any reason why s/he too should affirm that *p*. Won't liberals *mistakenly* vary the attitude they adopt in cases of this kind – just as the relativist mistakenly refuses to vary the attitude s/he adopts in cases of the kind earlier considered? I see no reason to suppose that they will. Surely, liberals don't suppose that every *apparent* reason-for-the-other is *ipso facto* a reason-for-them. (This is not required by their commitment to universalism and is incompatible with their commitment to fallibilism.) For one thing, they know that people are sometimes mistaken in thinking that they have genuine reasons for the attitudes which they have adopted, and so on these grounds – quite independent of questions here at issue – liberals will be disposed to *test* every alleged reason for affirmation or denial to see whether it provides *for them* a reason for affirmation or denial. And they will, in the course of such testing, surely recognize, when this is so, that the reasons of the other(s) provide *them* with no reasons for altering the attitudes which they have adopted.

Liberals are not vulnerable, then, at least in the same way, to a kind of mistaken inflexibility of attitude to which relativists are indeed subject (at least in worlds, like ours, in which aestheticism cannot be accepted as true). Liberals are not vulnerable, in other words, to a kind of irrationality to which relativists are indeed subject. They are invulnerable, as well, to mirror-image hyper-flexibility. The liberal attitude is therefore superior, with respect to rationality, to the relativistic attitude.

Since we can show, by an exactly parallel line of argumentation, that the liberal attitude is rationally preferable, as well, to the dogmatic attitude, we can show, in effect, that the liberal *attitude* is the uniquely *rational* attitude in the face of encountered disagreement.

II. THREE RESPONSES TO DISAGREEMENT

I want to consider now three familiar *responses* to disagreement – concession, coercion, and conversation, and to show that only the conversational response is, in situations of a certain kind, relevant to our concerns, consistent with the demands of rationality.

To the discovered existence of disagreement we might make a *concessive* response. We do so if we propose or accept a proposal to compromise in relation to the issue in dispute between us. So Alf and Beth make concessive responses to discovered disagreement about which path it would be right to take if, for instance, they flip a coin, agreeing to go to the left, as Alf wants, if the coin lands heads-up, and to go to the right, as Beth wants, if the coin lands tails-up.

To the discovered existence of disagreement we might make a *coercive response*. We do so if we attempt by force to compel our 'partners' to join us in that action which we think there is best reason to take. So Alf and Beth make coercive responses to discovered disagreement about which path it would be right to take if each attempts forcibly to compel the other to take the path which s/he prefers to take.

To the discovered existence of disagreement we might make a *conversational* response. We do so if we attempt conversationally to discover which of the parties to some issue in dispute has the best reasons for the position which s/he has adopted. So Alf and Beth make a conversational response to discovered disagreement about which path it would be right to take if they try conversationally to discover which of them can better support with evidence and argumentation the position on this issue which s/he has in fact adopted.

Now it might be thought that those who had adopted a liberal attitude could not, compatibly with their liberalism, respond coercively to the discovery of disagreement; that they could, compatibly with their liberalism, respond only conversationally (and in certain circumstances perhaps concessively). Indeed, this idea, although mistaken, is not implausible. Why would Alf and Beth want to coerce each other in relation to some issue about which each concedes, fallibilistically and universalistically, that s/he might be wrong and that the other might in turn be right?

It is important in this context to realize that what liberals like Alf and Beth reasonably want to *do* is what there would be best reason to do, so long as they do not, by doing this, unnecessarily disadvantage themselves. (Their purely cognitive interest in doing what there's best reason to do is moderated by the desire, by doing whatever they do indeed do, to achieve whatever non-cognitive goals they might have in mind.) So, if there is, for instance, best reason to take the right-hand path in order to return to the car, that is what both, if liberals, will want, *ceteris paribus*, to do; if there is best reason to take the left-hand path, then that is what both, if liberals, will want, *ceteris paribus*, to do.[11] But their commitment is not absolute to determine in this way what they ought to do. For they might, by proceeding in this way, make it impossible, or anyway very difficult, to achieve their non-cognitive goals – e.g. to return to the car as quickly as possible. Let's consider, in relation to this important point, a situation in which Alf and Beth might be willing to react coercively to discovered disagreement about the situation they face.

Imagine that one could at 'costs' C force the other to take the path which s/he initially wants to take. Imagine too that a conversational resolution of their dispute could be secured at costs C'. (Let C and C' be measured in the minutes it takes, respectively, for one to coerce the other or for them to conduct a dispute-resolving conversation.) Imagine, as well, that d is the measure (in minutes) of the differential between taking that path which there is best conversationally-discoverable reason to take and taking, instead, the longer of the two paths. Coercion will in this case be the rational response to disagreement if and only if $C + d$ is less than C', as it will sometimes surely be. For if this inequality does indeed obtain, then Alf and Beth will more quickly reach the car if one coerces the other and, as a result, they take the path that the coercer wanted to take than they will if Alf and Beth conversationally decide which path it would be better to take – and this will be so even if the path the coercer wants to take is the *longer* of the two. The rational response is, in other words, coercive, not conversational in this case.[12]

The liberal will, in a similar way and for the same kinds of reasons, sometimes prefer a concessive to a conversational response to the discovery of disagreement.

The liberal can sometimes rationally prefer coercion to conversation; s/he can sometimes prefer concession to conversation. Indeed, s/he can (consistently with the liberalism s/he exhibits) rationally adopt, as the precise circumstances dictate, any of the three basic responses to discovered disagreement. S/he will base the decision s/he makes on estimates of net expected benefit, choosing that response which s/he expects to maximize this quantity. S/he will converse only in those circumstances in which s/he expects by doing so to maximize this quantity.

I have in Section I argued that the liberal *attitude* to disagreement is the uniquely rational attitude. I seem to have shown, so far in this section, that there is no uniquely rational *response* to encountered disagreement. This is true, in a sense. Which of various possible responses is rational in a given case depends very much on the particularities of that case – on the relative costs associated with each response and, as well, on the likelihood that any given response will identify that course of action which there is best reason to take.[13] I nevertheless believe that there is a special class of cases, independently of interest, in which it can be shown that the conversational response is in fact, for liberals, the uniquely rational response to encountered disagreement.

III. SITUATIONS OF PURELY EPISTEMIC SIGNIFICANCE

We're all familiar with situations in which conduct depends on reaching a decision and waits on the decision-making process. Such *practically significant decisions* have to be implemented and are not really effected until they are implemented. (Alf and Beth accomplish nothing, really, until they reach the car.) With respect to decisions of this kind, procedural costs are therefore extremely important. If you haven't accomplished anything until you've returned to the car, then you haven't accomplished anything as long as you're still talking or fighting or otherwise trying to 'agree' on which way to go. In these situations, it's often at least as important to act quickly as it is to do the 'right' thing. If there really is very little difference in relative path-length, but very little time to decide which path to take, then it probably matters less which path Alf and Beth decide to take than it does how quickly they decide.

The 'decision costs' I mentioned in Section II above are therefore (and rightly) vitally important in the case of such practically significant decisions. (This is one of the reasons why it is, *in general*, hard to specify a uniquely rational response to disagreement (even when we assume that the disputants are both 'liberals').)

But it is nevertheless important not to lose sight of an obvious point, and one very much relevant to our concerns. *Some decisions have no practical significance*, or, more accurately, take place in circumstances which effectively nullify whatever practical significance they might in some other context have.

Again, examples are readily at hand. Alf and Beth might be scientists, investigating some natural or social phenomenon, and doing so purely for reasons of theoretical interest, and needing to base no form of practical conduct on the results of their investigations. They might disagree about how this phenomenon is best to be described or explained, Alf preferring one approach, Beth another and incompatible approach. But their interest is, as I have said, purely theoretical; their situation is one of 'purely epistemic significance', no matter of practical concern is affected by their disagreement, no cooperative project will be stymied by it, etc. Since there is no course of action which waits on their decision, 'decision-costs' are, in situations of this kind, of no significance (with respect to rationality).[14]

Now it is easy enough to show, with respect to situations of this kind, that the conversational response is the only response which is consistent with liberalism and is, therefore, the only rational response (a liberal attitude being, as shown in Section I above, the only rational attitude). To see this, we reason in two stages. We show, first, that liberals who encounter disagreement have reason to want to discover whether it is better to affirm, or, instead, to deny. We show, secondly, that liberals therefore have, in situations of purely epistemic significance, reason to respond to disagreement by means conversational rather than coercive or concessive in character. I take these points in turn.

What do those liberals *qua* liberals want to accomplish in situations in which they encounter disagreement? Let's look at it from the point of view of an individual Alf, who affirms that *p* and who encounters another individual Beth who denies that *p*. Since Alf is rational, he wants his attitude with respect to *p* to vary in accordance with reason – he wants to affirm that *p* if there is better reason to affirm and to deny that *p* if there is better reason to deny. Since Alf is a liberal he is *ipso facto* a universalist and therefore realizes that it is impossible that both he and Beth have good reasons for the attitudes they have adopted – if there is good reason to affirm, then there is, *ipso facto*, no good reason to deny (and vice versa): one of them must be wrong, given universalism, in thinking that s/he has a good reason for the attitude which s/he has antecedently adopted. Since Alf is a liberal he is *ipso facto* a fallibilist and therefore realizes that it cannot be ruled out that there is better reason to deny than there is to affirm that *p*: i.e. that *he* might be wrong to affirm that *p*. He will therefore reason as follows.

(1) I am obligated, as a rational being, to affirm if there is better reason to affirm and to deny if, instead, there is better reason to deny.

(2) If there is good reason to affirm, then there cannot be good reason to deny, and, on the contrary, if there is good reason to deny, then there cannot be good reason to affirm. It follows that we cannot both be right in adopting the attitudes which we have in fact adopted with respect to *p*.

(3) I might be wrong to think that there is good reason for the attitude which I have adopted with respect to *p*. In this case, there is good reason for the attitude which the other has adopted with respect to *p*.

(4) In order to fulfil my obligation as a rational being – see (1) above – I must discover for what reasons s/he has adopted the attitude which s/he has in fact adopted, and the relative merits of these reasons and my own.

(5) *Ceteris paribus*, I therefore need to attempt, conversationally, to resolve our disagreement. (Unless we *tell* each other why we have adopted the attitudes which we have antecedently adopted, I will not be able to tell which attitude there is *better* reason to adopt – nor will s/he.[15])

This argument is straightforward and we can, in situations of purely epistemic significance, ignore the *ceteris paribus* clause and conclude, without qualification, that *those who disagree need to talk*, in order, if possible, to resolve their disagreement.[16]

In any event, it is all well and good to say that those who disagree need to talk; that a conversational response to discovered disagreement is the uniquely liberal and therefore the uniquely rational response. It still remains to be specified how the resulting conversation ought indeed to be conducted. After all, there are many possibilities. Are Alf and Beth to debate one another in the manner of parliamentarians? Are they to shout abuse and hurl invective in the manner of fish-wives (and fish-husbands) brawling over the fence? Are they to flatter and cajole one another in the manner of travelling saleswo/men or of politicians seeking office?

IV. THE CONVERSATIONAL RESPONSE TO DISAGREEMENT

In order to establish the rationality of a conversational response to disagreement, it will be helpful to consider what's involved in making such a response. What do those individuals do who settle their disagreements conversationally? They talk with one another in accordance with the following rules.

(A) The First Rule of Respect

According to this rule, parties to the conversation cannot be silenced, but must be permitted to present the reasons which they have for adopting to the disputed proposition the attitudes which they have antecedently adopted. This rule imposes on participants in a liberal conversation an intellectual duty of non-interference.

(B) The Second Rule of Respect

This rule requires that each participant indicate why it is that, i.e. for what reasons, s/he has adopted to the disputed proposition the attitude which s/he has antecedently adopted. This rule, like the first, imposes on Alf and Beth an intellectual duty: in this case, the duty, given the opportunity which is secured by the first rule, to say on what grounds they have adopted the attitudes which they happen in fact to have adopted.

When individuals find themselves in disagreement they must, according to these two rules, each provide the other with an opportunity, and each accept the opportunity provided by the other, to say for what reasons s/he has adopted to the proposition in dispute between them the attitude which s/he has in fact adopted. Alf must permit Beth to say why it is that she denies that *p*; and Beth must not let the opportunity pass, she must not remain silent, but must, instead, specify her grounds for denying that *p*. Entirely symmetrically, Beth must permit Alf to say why it is that he affirms that *p*; and Alf must not let the opportunity pass, he must not remain silent, but must, instead, specify his grounds for affirming that *p*.

Of course, Alf and Beth will not get very far in *resolving* their disagreement if each is content simply to list the reasons s/he has for affirming or denying that *p*. Each will, of course, better understand the other ('He affirms that *p* because he believes that *q*'), but better understanding the other is not the *point* of the conversational exercise. Each wants to know what the other thinks not simply as a matter of idle intellectual curiosity (despite the purely epistemic significance of their dispute). Each wants to know, rather, whether the other has reasons that s/he too could acknowledge and, on the basis of which s/he should adopt to the disputed proposition an attitude which differs from that which s/he has antecedently adopted. Alf wants to know whether any of Beth's reasons is a good enough reason for denying that *p* to compel him too to deny that *p*. Beth wants to know whether any of Alf's reasons is a good enough reason for affirming that *p* to compel her too to affirm that *p*.

Alf and Beth therefore don't simply want to *discover* the other's reasons; they want, instead, to *compare* and *evaluate* the relative merits of these reasons. So it is not enough that each *listens* to what the other has to say. It is necessary, rather, that each *responds* in a certain way to what the other has to say. And this brings us, indeed, to:

(C) The Rule of Responsibility

This rule proscribes a null response to contributions made to a conversation about the relative merits of competing attitudes to *p*. Suppose that Alf specifies as a reason for affirming that *p* some consideration *R* which seems to show that it is in fact correct to affirm that *p*. The rule of responsibility requires of Beth that she respond to this contribution, *either* by accepting that *R is* indeed a reason for affirming that *p*, and so herself affirming that *p*, or by introducing some consideration R^* which seems to show that the consideration *R* does not in fact provide adequate grounds for affirming that *p*.

Beth might claim, for instance, that the consideration *R* does not satisfy one or more of the requirements set out at (D)-(F) below – requirements which place certain substantive limits on what *counts as a reason* for affirming or denying any disputed proposition. Beth would in this way satisfy the 'put up' clause of this rule, which requires, really, simply that disputants 'Put up or shut up' in the face of opposition to their points of view.

Or Beth might be unable in this way to meet with counter-arguments those considerations presented by Alf as grounds for affirming that *p*. She must, in this case, join him in affirming that *p*, and she will, if she does do so, satisfy the 'shut up' clause of this rule.[17]

This rule, like the others so far considered, is a duty-imposing rule. The rule of responsibility imposes on conversationalists the duty to respond to the contributions of their interlocutors. Should they be unable to do so, then they must abandon as rationally unsupportable the attitude they have antecedently adopted, adopting, instead, that of their interlocutors. (Obviously, this rule is crucial in insuring that conversation results, if possible, in *convergence*, i.e. in the *resolution* of disagreement. If it were required of disputants merely that they list for one another their reasons for affirmation or for denial, then there would be no *impetus* to convergence of opinion. It is only insofar as disputants are required either to respond to criticism or to acknowledge its cogency that convergence can indeed be engineered.)

The rules next to be considered identify those considerations which participants in a liberal conversation can, in conformity with their duties of participation, rightly introduce as grounds for affirmation or denial of disputed propositions. These *reasons-defining* rules in effect specify grounds for excluding or rejecting claims which otherwise rule-abiding participants in a liberal conversation might otherwise attempt to introduce. Claims which are proscribed by these rules will not be recognized as legitimate contributions to discussion. If Alf introduces as grounds for affirming that *p* some consideration *R* which does not satisfy requirements laid down by one of these rules, then *R* need not be, indeed should not be, recognized as a legitimate contribution to the discussion and Beth can, to satisfy the demands of the rule of responsibility, cite as grounds for rejecting *R* (and therefore for continuing to deny that *p*) the fact that *R* in fact fails to satisfy these requirements, which are specified, in particular, by the rule of relevance.

(D) The Rule of Relevance

This is the most fundamental of three rules which are reasons-defining (rather than duty-imposing), and it requires, in particular, that no consideration *R* be recognized as a reason for affirming or denying some disputed proposition *p* unless *R* is in some logically respectable way *relevant* to *p*. Considerations which are (in some broadly logical sense) irrelevant to the issue in dispute cannot legitimately be recognized as grounds for preference in relation to this issue.[18] So, if Alf introduces considerations *R* as grounds for affirming that *p* (or for rejecting some consideration *R′* introduced by Beth as grounds for denying that *p*), then it is required by this rule that these considerations have some bearing on the truth or falsity of the claim in question – viz. that *R* imply the truth of *p* (or the falsity of *R′*), or that *R* make it more likely that *p* is true (or that R′ is false), etc. So Beth could, to satisfy the obligation of responsible argumentation, reject *R* on this basis; she could, in other words, 'put up' and could do so by adverting to the irrelevant character, which disqualifies it as a genuine reason for affirmation, of Alf's contribution.

(E) The Rule of Neutrality

According to this rule, it does not count as a reason for affirming or denying some disputed proposition that that proposition is compatible or incompatible with some 'sacred text' or with the saying of some 'holy person'.

Suppose that Beth introduces as grounds for denying that *p* the claim *R*: that *p* is incompatible with the views of some distinguished figure or with the content of some respected text. Although Beth here *seems* to conform to the duty imposed by the second rule of respect – although she *seems*, in other words, to provide *grounds* for denying that *p* – according to this rule, her contribution cannot be recognized as legitimate. To suppose that incompatibility with the views of some person or with the contents of some text is by itself grounds for denying that *p* is to attribute to the person or the text in question a kind of incontestable authority that is incompatible with the fallibilism on which the liberal attitude to disagreement is in part founded.

Since Alf must argue 'responsibly', he could, in this case, discharge this duty by drawing attention to the non-neutral character of Beth's contribution, which disqualifies it as a genuine *reason* for denial.

(F) The Rule of Publicity

According to this rule, no consideration *R* counts as a reason for affirming or denying some disputed proposition *p* unless each of the disputants about *p* can *independently* affirm that *p*. Considerations which one party can but which the other party cannot affirm cannot legitimately be recognized as grounds for affirmation or denial in relation to this issue. Reasons must be *publicly affirmable*; they must not presuppose or imply the authority of any of the parties to the dispute (or of anyone else, for that matter).[19]

Suppose that Alf introduces as grounds for denying that *p* the claim *R*: that he has had some experience, which Beth is not in a position to have, which convinces him that *p* is true. Although Alf here *seems* to conform to the duty imposed by the second rule of respect – although he *seems*, in other words, to provide *grounds* for affirming that *p* – according to this rule, his contribution cannot be recognized as legitimate. To suppose that experiences which Alf has had but which Beth cannot have provide both with grounds for affirming that *p* is to attribute to Alf's claims about this experience a kind of incontestable authority that is incompatible with the fallibilism on which the liberal attitude to disagreement is in part founded.

Since Beth must argue 'responsibly', she could, in this case, discharge this duty by drawing attention to the purely private character of Alf's contribution, which disqualifies it as a genuine *reason* for denial.

These three reasons-defining rules specify conditions which must be satisfied if considerations introduced by the participants in liberal conversations are to be recognized as legitimate. Considerations which fail to satisfy one or more of these conditions cannot be recognized as legitimate contributions to such conversations (and the duty of participants to 'put up or shut up' can be discharged by pointing out that such considerations do in fact fail to satisfy one or more of these conditions).

We have now specified duty-imposing and reasons-defining rules of liberal conversation. These rules in effect generate a discursive practice of a particular kind, in which disputants about *p* are permitted and obligated to offer reasons in defense of their attitudes with respect to *p* and to offer reasons in defense of their reasons in defense of their attitudes with respect to *p*, and so on – acknowledging, when they can no longer undermine the considerations introduced by their opponents, that their opponents have better reason than they have for the attitudes they have adopted. At this point disagreement is, at least *pro tem*, resolved and convergence achieved – each adopts to the disputed proposition the attitude which there is best reason to adopt and each agrees with the other(s) about *which* attitude there is indeed best reason to adopt. Their obligations as rational beings are in this way discharged.

It remains, of course, to *justify* the imposition of these rules on conversations aimed at the resolution of disputes, an exercise to which I now turn.

V. JUSTIFYING THE RULES OF CONVERSATIONAL GOOD CONDUCT

How are we to justify the imposition, on discussions aimed at the resolution of disputes, of the six rules of liberal conversation? There are, in fact, two rather different ways in which we might proceed, and it will be instructive before continuing briefly to distinguish these two approaches. I begin by appealing to a context in which this distinction is already familiar.

Within the context of *ethical* reasoning there has for many years been recognized a distinction between two main frameworks for justifying the imposition of moral rules.

We might say, within a broadly *teleological* framework, that an agent *A* was, in a situation Σ, *right* to perform an action Ø. If we did so, we would mean, in particular, that *A*'s Øing in Σ was, in its *effects* and with respect to some goal which moral agents ought to strive to realize, better than any alternative action would in that situation have been. An action's right-making characteristics are, on this account, its effects with respect to the achievement of some independently specified goal of moral conduct (e.g. the maximization of total happiness). Actions are right, on this account, to the extent to which they are productive of good effects. And, derivatively, though for our purposes more importantly, we might say that we are justified in imposing a *rule* which requires Øing in Σ if and only if behavior in conformity with such a rule is conducive to the achievement of some independently specifiable goal of moral conduct.

We might say, within a broadly *deontological* framework, that an agent *A* was, in a situation Σ, *right* to perform an action Ø. If we did so, we would mean, in particular, that *A*'s Øing in Σ was, in its intrinsic *qualities*, better expressive than any alternative action would in that situation have been of the individual's *commitments* as a moral agent. An action's right-making characteristics are, on this account, its intrinsic qualities vis à vis the commitments which define the agent's circumstances. Actions are right, on this account, to the extent to which they are, not in their more remote effects, but instead in their own intrinsic qualities, expressive of the agent's commitments. And, derivatively, though for our purposes more importantly, we might say that we are justified in imposing a *rule* which requires Øing in Σ if and only if behavior in conformity with such a rule is expressive of the individual's commitments as a moral agent.

So, while it may be possible teleologically to demonstrate that *A* is right in Øing, it may be possible as well to show non-teleologically that s/he is right in Øing. Both of these broad general strategies are widely recognized, though not of course universally adopted, by ethical theorists.

It is my intention here to exhibit a broadly 'deontological' alternative to the commonly accepted teleological framework for justifying such rules for the resolution of disputed issues as I have specified. I hope to show, in other words, that liberal conversationalists have non-teleological reason to be guided by these rules. I hope to show, more specifically, that conformity with these rules is better expressive than non-conformity would be of the conversationalists' commitments as rational agents facing encountered disagreement, and that it is therefore rational deontologically, expressively, that they conform to these rules – not because of any anticipated benefits (though these may result), but, instead, because of the intrinsic qualities, relative to their commitments, of conversation in conformity with these rules. Just as the deontological moral theorist demonstrates the rightness of giving by showing that giving better than withholding expresses the agent's commitment to give, so too does the deontological epistemic theorist demonstrate the rationality of conformity with the rules of some method by showing that conformity rather than non-conformity better expresses the agent's commitments as a rational being.

If we are deontologically to justify imposing on liberal conversationalists the six rules previously specified, it will be necessary, first, to characterize their commitments, and, secondly, to demonstrate the expressivity of these commitments of behavior in conformity with each of these rules. I take these points in turn.

What are the commitments of liberal conversationalists? They are readily enough characterized and have, indeed, already been identified. Liberal conversationalists are committed (and it is rational that they be so committed) to forming their attitudes (with respect to disputed propositions) in accordance with the demands of reason, or, in other words, to affirming if there is better reason to affirm, to denying if, instead, there is better reason to deny. Notice, also, that *liberal* conversationalists are committed to a universalistic approach to reasoning (what is a reason-for-one is *ipso facto* a reason-for-all) and to a fallibilistic approach to experience (there is, except in trivial cases, no such thing as incontestable first-person authority).

Is conduct in conformity with the rules proposed better expressive than conduct not in conformity with these rules of the commitment of liberal conversationalists? Transparently, conforming behavior better expresses than does non-conforming behavior the commitment of liberal conversationalists. Let us examine, in turn, each of the six proposed rules.

(A) The First Rule of Respect

'You must not silence those who disagree with you!' – this is what this rule means for liberal conversationalists who face a situation of discovered disagreement. Consider this rule from Alf's point of view.

Alf is committed, as a rational being, to discovering whether it would be better from the point of view of rationality to affirm or instead to deny that *p*. Alf therefore wants to discover why it is that Beth denies that *p*. Until he does so, he cannot be sure that his reasons for affirmation are better than whatever reasons for denial she might happen to be aware of. He therefore has good reason to object to any attempt on the part of others to silence Beth, and good reason to refrain himself from any such attempt. For were Beth to be silenced, Alf would not at least from her discover her reasons for denying that *p*, and would *a fortiori* be unable to discover whether these reasons are cogent and weighty enough to outweigh whatever reasons he has for affirming that *p*.

Alf can be thought of as facing a choice: to silence Beth or instead to hear her out. He brings to this choice a commitment: a commitment to adopt to disputed propositions whichever attitude there is better reason to adopt. If he silences Beth this commitment cannot be honored – for he will not know why she denies and therefore will not know whether her grounds for denial are better than his for affirmation. Forbearance therefore better expresses his commitment than would any attempt to prevent Beth from speaking. But forbearance is required by the first rule of respect. Behavior in conformity with this rule therefore better expresses than behavior not in conformity with it Alf's commitment as a rational being, and the imposition of this rule is therefore justified, deontologically.

(B) The Second Rule of Respect

'You must not remain silent in the face of disagreement!' – this is what this rule means for liberal conversationalists who face a situation of discovered disagreement. Consider this rule from Beth's point of view.

Beth is committed, as a rational being, to discovering whether it would be better from the point of view of rationality to affirm or instead to deny that *p*. Beth therefore wants to discover whether any of Alf's reasons for affirming that *p* is a good enough reason for affirming that *p* to compel her own affirmation of *p*. But Beth cannot tell how good are Alf's reasons for affirming that *p* until she 'sees' how well he is able to defend his affirmation of *p* against those considerations which persuade her, at least antecedently, instead to deny that *p*. But Alf will have no opportunity to defend his affirmation unless he is told by Beth (he cannot read her mind) on what grounds she instead denies.

Beth can be thought of as facing a choice: to remain silent or, instead, to participate in the attempt, conversationally, to resolve disagreement. She brings to this choice a commitment: a commitment to adopt to disputed propositions whichever attitude there is better reason to adopt. If she remains silent this commitment cannot be honored – for she will not know, since she has given him no opportunity to show this, whether Alf has good reasons for rejecting as infirm her reasons for denying that *p*. Participation therefore better expresses her commitment than would any attempt to remain silent in the face of Alf's affirmation of *p*. But participation is required by the second rule of respect. Behavior in conformity with this rule therefore better expresses than behavior not in conformity with it Beth's commitment as a rational being, and the imposition of this rule is therefore justified, deontologically.

(C) The Rule of Responsibility

'Put up or shut up!' – this is what this rule means for liberal conversationalists who face a situation of discovered disagreement. Consider this rule from Alf's point of view.

Alf is committed, as a rational being, to discovering whether it would be better from the point of view of rationality to affirm or instead to deny that *p*. Because he wants to discover, in particular, whether he has any reason to abandon his affirmation of *p*, Alf wants to discover whether or not any of Beth's grounds for denying that *p* in fact constitutes a good reason for denying that *p*. He therefore has reason, with respect to each such consideration *R* for denying that *p*, to try to discover whether or not there are considerations R^* which would warrant rejecting the consideration *R* as grounds for denying that *p*. For unless he can establish with respect to a given consideration *R* that this consideration is not, in view of other considerations R^*, a good (enough) reason for denying that *p*, Alf has reason to believe that he is wrong to affirm that *p*. For the consideration *R* would, unless undermined or outweighed by some other consideration R^*, constitute a good reason for denying that *p*.

Alf can be thought of as facing a choice: to ignore, despite being confronted by, Beth's evidence and arguments or, instead, to argue responsibly, rejecting only that evidence and those arguments which there are good reasons to reject. He brings to this choice a commitment: a commitment to adopt to disputed propositions whichever attitude there is better reason to adopt. If he ignores Beth's contributions this commitment cannot be honored – for he will in this case reject without good reason to do so considerations which might constitute good grounds for denying that *p*. Responsible argumentation therefore better expresses his commitment than would any attempt simply to ignore Beth's ideas. But responsible argumentation is required by the rule of responsibility. Behavior in conformity with this rule therefore better expresses than behavior not in conformity with it Alf's commitment as a rational being, and the imposition of this rule is therefore justified, deontologically.

(D) The Rule of Relevance

According to this rule, no consideration *R* counts as a reason for affirming or denying some disputed proposition *p* unless *R* is in some logically respectable way *relevant* to *p*. Considerations which are (in some broadly logical sense) irrelevant to the issue in dispute cannot legitimately be recognized as grounds for preference in relation to this issue. That this is so is, it seems to me, inherent in the idea that one proposition is a *reason* for affirmation or denial of some other proposition. Those who are committed to adopting that attitude which there is better *reason* to adopt are therefore *ipso facto* committed to rejecting any consideration introduced by their opponents which stands in no logically respectable relation of relevance to the issue in dispute between them.

The point is, clearly, deontological in character. I do not claim that no attitude grounded in logically irrelevant claims could possibly be a correct attitude or that grounding in such claims is necessary if our attitude is to conform to the 'facts'. We may sometimes be persuaded by irrelevancies to affirm what it would (otherwise) be right to affirm. My point is, simply, that affirmation on the basis of relevant considerations better expresses than would affirmation on the basis of considerations which are not relevant the commitment of the rational agent – to affirm if and only if there is better *reason* to affirm.

(E) The Rule of Neutrality

According to this rule, no consideration *R* counts as a reason for affirming or denying some disputed proposition *p* if it involves as appeal to the authority of some text or person. Consider this rule from Beth's point of view.

Beth is committed, as a rational being, to discovering whether it would be better from the point of view of rationality to affirm or instead to deny that *p*. Because she is committed to a fallibilistic understanding of human experience, Beth wants to affirm or deny that *p* only on the basis of those considerations which she can herself independently affirm. Suppose that Alf recommends affirming that *p* because affirmation (but not denial) is consistent with the sayings of some person or with the contents of some text. There are two cases: (a) Beth can independently affirm the sayings or the text at least insofar as they are relevant to the truth or falsity of *p*. In this case, Beth will join Alf in affirming that *p*, but will do so, not because of the *authority*, but, instead, because of the veracity, independently confirmed by her, of the person or the text. (b) Beth cannot independently affirm the sayings or the text insofar as they are relevant to the truth or falsity of *p*. In this case, Beth will refuse to join Alf in affirming that *p*, since the only grounds he provides for affirming that *p* are considerations which she cannot herself independently affirm. To accept these grounds would amount, in effect, to accepting as incontestably authoritative the person or the text in question and this Beth cannot do, consistently with her commitment to fallibilism.

Beth can be thought of as facing a choice: to accept the authority of persons or texts despite her inability independently to confirm their veracity, or, instead, to accept as legitimate grounds for affirming (or for denying) that *p* only those considerations which she can herself independently affirm. She brings to this choice a commitment: a commitment to a fallibilistic understanding of human experience. If she accepts the authority of texts or of persons, then this commitment cannot be honored – for she will in this case treat such texts or persons as if they were incontestably authoritative. Argumentation in accordance with the rule of neutrality therefore better expresses her commitment than would any appeal to the authority of such texts or persons, and the imposition of this rule is therefore justified, deontologically.

(F) The Rule of Publicity

According to this rule, no consideration *R* counts as a reason for affirming or denying some disputed proposition *p* unless disputants about *p* can independently affirm that *R*. Considerations which one party can but which the other party cannot affirm cannot legitimately be recognized as grounds for preference in relation to this issue. Consider this rule from Alf's point of view.

Alf is committed, as a rational being, to discovering whether it would be better from the point of view of rationality to affirm or instead to deny that *p*. Because he is committed to a fallibilistic understanding of human experience, Alf wants to affirm or deny that *p* only on the basis of those considerations which both he and Beth can independently affirm. Suppose that considerations *R* constitute grounds for Alf to affirm that *p*. Suppose he discovers that Beth cannot join him in affirming that *R*. He will in this case concede that *p* cannot *on these grounds* be affirmed, for he cannot, consistently with his commitment to fallibilism, insist, in the face of Beth's rejection of *R*, that he is right to affirm that *R*, and, therefore, on this basis to affirm that *p*. To do so would be, in effect, to insist on his incontestable first-person authority with respect to *R*, and this is incompatible with a commitment to fallibilism.

Alf can be thought of as facing a choice: to insist despite Beth's dissent on the cogency of his reasons for affirming that *p*, or, instead, to accept as legitimate grounds for affirming (or for denying) that *p* only those considerations which (both he and) Beth can independently affirm. He brings to this choice a commitment: a commitment to a fallibilistic understanding of human experience. If he insists on the cogency despite dissent of the considerations which move him to affirm that *p*, then this commitment cannot be honored – for he will in this case treat these considerations as if they were incontestably authoritative. Argumentation in accordance with the rule of publicity therefore better expresses his commitment than would any appeal to the authority of his own experience, and the imposition of this rule is therefore justified, deontologically.

I claim to have shown, then, that the six rules of liberal conversation can be justified deontologically, and, in particular, that conforming behavior is better expressive than non-conforming behavior of the fundamental epistemic commitments of liberal conversationalists – viz. to adopt to disputed propositions that attitude which there is better reason to adopt, and to do so mindful of the demands of fallibilism and of universalism.

The liberal attitude to encountered disagreement is the uniquely rational attitude. Those who adopt such an attitude have reason, deontologically, to submit themselves, in situations of purely epistemic significance, to the six rules of liberal conversation herein specified, and conversations conducted in accordance with these rules are ones in which the liberal's commitments as a rational being are necessarily honored. Disputants have, then, very good reason indeed to conduct themselves in accordance with these rules.

* * *

The rationality of science consists, then, in just this: – Science is rational because or insofar its disagreements are resolved in accordance with the rules of liberal conversation. And it is rational to follow such rules, not because doing so enables one to realize some aim, but, instead, because doing so enables one to express one's commitments as a rational agent, committed to a liberal attitude, in the fact of encountered disagreement.[20]

Department of Philosophy
University of New England

NOTES

[1] For some references, see my papers 'The Aimless Rationality of Science', *International Studies in the Philosophy of Science*, vol. 4 (1990) and 'Adjudication as an Epistemological Concept', *Synthese*, vol. 79 (1989).

[2] See John Harsanyi, 'Advances in Understanding Rational Behaviour', in J. Elster, ed., *Rational Choice* (Oxford, 1986: Basil Blackwell).

[3] My primary sources of inspiration are Bruce Ackerman, *Social Justice in the Liberal State* (New Haven, 1980: Yale University Press) and John Rawls, *A Theory of Justice* (Oxford, 1973: Oxford University Press). Within the same general framework adopted here in relation to problems of scientific methodology, I consider some problems of political theory in my papers 'Relativism and Reflective Equilibrium', *The Monist*, vol. 71 (1988) and 'Ethical Pluralism and the Role of Opposition in Democratic Politics', *The Monist*, vol. 73 (1990).

[4] Reasons are sometimes said to be 'universalizable', in the sense that what is rightly said to be a reason for one is *ipso facto* rightly said to be a reason for *all* to think or to act in some particular way. This claim is accepted by 'universalists' and rejected by relativists.

[5] Suppose that Alf prefers chocolate to vanilla ice-cream on the grounds that chocolate *tastes better to him* than vanilla does. Suppose that Beth prefers vanilla to chocolate on the grounds that vanilla *tastes better to her* than chocolate does. It seems that each of them has an entirely adequate reason for the preference s/he has formed; that the reason each of them has is a reason *for that person but not for the other*; and, therefore, that there is in the nature of such a 'dispute' nothing that either of them could say to the other that might provide a reason why s/he should alter the preference which s/he has formed. A relativistic attitude seems an altogether sensible one in a case of this kind.

[6] The aesthete believes, in other words, that there is no disputed proposition about which either 1 or 2 above is true. In every case, it is, instead, 3 that is true.

[7] We distinguish in modal terms the dogmatist and the 'ordinary', fallibilistic believer of p. Even the fallibilist believes, of any proposition q incompatible with some proposition p which s/he believes, that q is false. But s/he does not (at least in general) believe, as the dogmatist does, that such a proposition q could not possibly be true.

[8] Suppose that Alf claims to be in pain and that Beth denies that this is so. Surely, Alf would be 'within his rights' to reject any considerations introduced by Beth which are intended to support her claim that he is not in fact in pain – and to do so solely on the grounds that the falsity of these considerations is revealed in their very incompatibility with his incontestably authoritative first-person experience of his pain. A dogmatic attitude seems an altogether sensible one in a case of this kind.

[9] Important influences on my description of the liberal attitude include Jacob Bronowski, *Science and Human Values* (London, 1961: Hutchinson); Karl Popper, *The Open Society and Its Enemies* (London, 1945: Hutchinson); Bruce Ackerman, *Social Justice in the Liberal State*, *op. cit.*; and Robert Merton, 'Science and Democratic Social Structure', in *Social Theory and Social Structure*, rev. edn. (New York, 1957: The Free Press).

[10] I do not believe that this definition begs questions about the rationality of relativism and/or dogmatism. While the definition does imply that the aesthete and the authoritarian are not rational in their attitudes to encountered disagreement, this, surely, is unobjectionable in view of the manifest falsity of these overblown doctrines. Certainly, the relativist is not branded *ipso facto* as an irrational individual. Where there is a relativity of reasons the relativist will vary the attitude s/he adopts entirely in accordance with reasons which s/he can recognize as reasons, and will thus satisfy the demands of the definition. In a similar way, a dogmatic attitude need not be *ipso facto* an irrational attitude. Where there is incontestable first-person authority the dogmatist will never be presented with anything other than specious 'reasons' for changing the attitude s/he has adopted and therefore cannot be charged with irrationality if s/he refuses on these grounds to alter the attitude s/he has adopted.

[11] As will become clear, the *ceteris paribus* clause is to be read as alluding, primarily, to so-called 'decision-costs'. It cannot be rational, even for a liberal, to spend so much time engaged in conversation that there is too little time remaining after conversation has been completed actually to implement the course of action decided on.

[12] These considerations point to a certain ambiguity in our notion of rationality. There are (at least) two senses which need to be distinguished. Taking the left-hand path is *epistemically* rational if there is reason to believe that this path is the shorter of the two. Taking the left-hand path is *instrumentally* rational if by doing so one can more quickly return to the car. Because it

can be quite time-consuming to discover which action is epistemically rational, that action is not the one which it is always instrumentally rational to take (at least if doing so requires first determining which action is the epistemically rational one).

[13] Assume, realistically, that there is no correlation between being that course of action which there is best reason to undertake and being that course of action which is stochastically or combatively selected (e.g. by flipping coins or by trial-by-combat). In this case, stochastic and combative procedures will select the best course of action 50% and conversational procedures will select the best course of action 100% of the time.

[14] The parenthetical qualification is necessary. There are, undoubtedly, disagreements of purely epistemic significance which have associated with them such high decision-costs that a conversational resolution could not *reasonably* be contemplated.

[15] Conversation is necessary, in part, because (or to the extent to which) 'mind-reading' is impossible. I can't tell whose reasons are better unless I know what your reasons are. I can't know what your reasons are unless you tell me. I can't know whose reasons are better unless you tell me what your reasons are. This is why a specifically *conversational* approach is demanded.

[16] Notice, by the way, the sense in which they 'need' to do this. It is not, at least in situations of purely epistemic significance, that they need to talk in order, by resolving their disagreement, to facilitate or at least to remove some impediment to some conjoint practical activity. In situations of purely epistemic significance, there is nothing that Alf and Beth need to *do* that they can't do until they resolve their disagreement – and so that's not why they need to resolve it. They need to resolve it, in situations of this kind, purely as an expression of their nature as rational beings, purely in order that they might know – what is by their discovery of disagreement thrown into doubt – what there is better reason to *think*. Until they resolve their disagreement, they simply don't know what to think about *p*, the proposition in dispute between them, and rational beings want to know, at least when the question arises, which attitude to some proposition there is better reason to adopt – and they want to know this simply as rational beings.

[17] Liberal conversation thus involves that same 'method of constrained silence' as is recommended in the context of political disputation by Bruce Ackerman, *Social Justice in the Liberal State*, p. 10. Those who have, in their defense of some attitude with respect to *p*, been reduced to silence by their opponents cannot any longer legitimately claim that their attitude is the right one.

[18] I write here in terms which are very vague. While there are (nearly perfectly) clear cases of relevance (of a is *P*-ish but not *Q*-ish to All *P*s are *Q*s) and irrelevance (of a is *R*-ish to All *P*s are *Q*s), purely deductive relations of these kinds need not exhaust the analysis of the relations of relevance and irrelevance. It is not herein precluded that there are relations of relevance and irrelevance which cannot be characterized in purely deductive terms.

[19] I write here in terms which are very vague. While there are clear enough cases of publicly and of purely privately evaluable claims, many other cases are much more problematic. Whether a consideration *R* is legitimately to be considered public in the relevant sense is a question on which reasonable people might reasonably differ.

[20] I am grateful to John Watkins and Jozef Misiek for the opportunity of participate in the Krakow conference of June 1989, and to the Latsis Foundation for its generous sponsorship. I am grateful to my Polish hosts for their wonderful hospitality, in the midst of events which were, to say the least, exciting. I am grateful to Peter Forrest, John Watkins, Jerry Gaus, Stephanie Lawson, and Jeff Malpas for help of various kinds.

D. M. ARMSTRONG

WHAT MAKES INDUCTION RATIONAL?*

1. INTRODUCTION

In this paper I put forward what I think is a new approach to the problem of induction. I sketched the approach in brief sections of a book published in 1983 (chap. 4, sec. 5 and chap. 6, sec. 7). The same idea had occurred to the English philosopher John Foster and he presented it in a paper at about the same time (1983).

The apparent coincidence was not much of a coincidence. There was a common cause. As we shall see, the arguments rest upon two pillars. The first pillar is that good scientific inference often takes what C. S. Pierce called an "abductive" form (Buchler 1940, chap. 11). Gilbert Harman characterized such arguments by introducing a phrase that is now a catch phrase: *inference to the best explanation*. The importance of such inferences in natural science has become more and more evident to contemporary philosophers. Inference to the best explanation was in the mind of both Foster and myself.

The second pillar is that the laws of nature are not mere regularities in the behaviour of things, but, rather, involve what Foster calls "objective natural necessities." Both Foster and I reject the "weak" Humean or mere regularity conception of laws, in favour of "strong" laws.

Once one is sensitized to the importance of the notion of inference to the best explanation *and* one has come to accept the notion of strong laws of nature, it is not unlikely that one will hit upon, and be attracted to, the Foster-Armstrong approach to induction. For the core idea is very simple: observed regularities are best explained by hypotheses of strong laws, hypotheses which in turn entail conclusions about the unobserved.

2. REJECTION OF INDUCTIVE SCEPTICISM

It will be desirable as a preliminary to indicate what I take the argument to show and not to show. I do *not* think that what I say will refute inductive scepticism. This does not worry me much because I agree with David Stove (1986) that for a philosopher to embrace inductive scepticism is to embrace a position that is irrational from the outset.

J. Misiek (Ed.), The Problem of Rationality in Science and its Philosophy, 45–54.

Kluwer Academic Publishers.

All of us make a huge number of inductive inferences and, as a result, take ourselves to know very many truths. Hume gives three good examples. We have ourselves to have *discovered by experience* that bread nourishes, water suffocates and fire burns.

These beliefs of ours belong to a still larger class of beliefs which I call *Moorean* beliefs. The reference, of course, is to G.E. Moore's defence of common sense (1925). Our assurance of the truth of such beliefs is so great that we hardly ever think of them or assert them. Indeed, it is almost embarrassing so to do. We are quite certain of these things or, as we are prepared to say, they are quite certainly true. I may add that that the class of Moorean beliefs is a mere subclass of those beliefs which are sometimes called "folk theories." There are many instances of false folk theories.

Our Moorean beliefs are such that there is something irrational in giving any reasons in believing them, except perhaps other Moorean beliefs. Giving "external" reasons for the truth of such beliefs, for instance, of the existence of a good God who would not deceive us, is silly because the external reasons are very much less certain than the Moorean corpus which is being endorsed.

It is equally irrational to advance reasons *against* a Moorean belief. (A possible exception is the use of one Moorean belief to cast doubt upon another Moorean belief.) To give external reasons to doubt a Moorean belief is irrational because, once again, the propositions advanced as reasons are very much less certain than the Moorean corpus.

Philosophical reasoning which seeks to make us sceptical about some or all of the Moorean corpus is a clear case of bringing external reasons to bear upon the corpus. Given the known difficulty of obtaining *knowledge* in philosophy, it will be indefinitely more likely that there is a flaw somewhere in the philosophical argument than that the Moorean corpus contains a mistake. The primitive certainty of having had breakfast this morning completely outweighs fine-spun arguments to show time unreal, flawless though the latter may seem to their propounders.

In the case of induction, we have Hume's beautiful argument to show that it is not rational to argue to the nature of the unobserved on the basis of the observed. But included in the Moorean corpus are both the general proposition that we can often learn about the unobserved on the basis of the observed, and innumerable specific propositions, for instance, the ones about bread, water and fire already mentioned. Just consider the thought-experiment of bracketing off the inductive principle and specific beliefs arrived at by the use of induction from the rest of Moorean corpus. How much would there be left? No more than this, apparently: current awareness of various things having, or seeming to have, various sensible properties and relations, together with some memory of past perceptions of a similar sort. Solipsism of the present moment is not far away.

So it must be rational to back our common-sense view that experience can teach us, and has taught us, much about the unobserved, against the philosophical arguments of the inductive sceptic.

What, then, may we hope that a justification of induction will do? This, I think. Ever since Hume, inside every philosopher who has attended to Hume's argument, there is an inductive sceptic. After the argument has been laid out, it becomes very hard to see *exactly* why inductive arguments are good arguments. I have just argued that we would be utterly foolish to deny that such arguments *are* good, but we can (and do) still wonder why they are. Perhaps this is just an "irritable groping after fact and reason" where no further fact or reason exists. Perhaps experience *teaches*, really teaches, but that in the end is all there is to be said. But, then again, perhaps a structure of reasoning can be laid out which will more fully reconcile us (the philosophers) to the rationality of induction.

What follows is an attempted justification of induction in that rather minimal sense.

3. INFERENCE TO THE BEST EXPLANATION

There is very primitive picture of scientific inference which, except for any mathematics or other deductive process that may be involved, makes it all inductive. *All* non-deductive inference in the natural sciences has the following form. The observed *F*s are all *G*s, so it may be inferred (though not deductively inferred) that all *F*s, and in particular the unobserved *F*s, are *G*s.

Many years ago, more or less following Berkeley, I argued against the Representative theory of perception in the following way. The inference from the immediately perceived sense-impressions to external physical objects which are their causes must either be deductive or inductive (by Reason or Sense, as Berkeley put it). Clearly it is not deductive. But neither is it inductive. For, by hypothesis, we have no direct *experience* of physical objects, still less to physical objects giving rise to sense-impressions. So, if we start from the truth of the Representative theory, we can have no good reasons to believe that the physical word exists.

It was J. J. Smart who pointed out to me what a bad argument this is. Why should not an upholder of the Representative theory claim that our belief in the existence of a physical world is justified not inductively but rather as a good explanation for the particular nature of the flow of our sense-impressions? If we *hypothesize* that there is a continuing physical world lying beyond our perceptions, and causing them, then the relative regularity and order of perceptions will become intelligible. Here we have a putative case of what came to be called "inference to the best explanation."

What general principles, if any, govern this sort of inference? One suggestion for an answer grows out of the work of Hans Reichenbach (1956), work developed more recently by Wesley Salmon (1978). This is the Principle of the Common Cause. Suppose that there is a certain conjunction of events, A, B, C, … . The conjunction may be a *mere* conjunction, or the conjunction may spring from a common cause. In a great many cases, the hypothesis of mere conjunction will be *less probable* than the hypothesis that the events (which may be tokens or types) have a common cause. So it will be reasonable to postulate a common cause. We find the principle at work in the thought of that philosopher of science, Ian Fleming's James Bond: "Once is happenstance, twice is coincidence, but three times is enemy action." Reichenbach puts it thus: "If an improbable coincidence has occurred, there must exist a common cause" (1956, p. 157).

Smart (1982) has pointed out that Reichenbach's principle is well suited to the defence of scientific realism, in particular where a *single* entity is postulated to stand behind a *number* of more or less observable phenomena. But Smart also notes something else, something more immediately relevant to our present theme. It is that there seems to be no reason to restrict the Principle to *causes*, strictly so called. Suppose that we have two hypotheses about a certain conjunction of phenomena. The first is that the conjunction is a bare conjunction: bare coincidence. The second hypothesis is that the existence or nature of each conjunct can be explained, causally or in some other way, as flowing from the same, single principle. In general, and other things being equal, we will think that this second hypothesis is more likely to be true.

It is this extended principle, the Principle of the Common Principle as it were, that will now be appealed to as a way of justifying induction, where "justifying" has no more force than the limited sense proposed above in Part 2. It will involve arguing that classical or extrapolative induction (all observed *F*s are *G*s, so it is likely that all *F*s are *G*s) is simply a particular case of abduction, where the abductive argument proceeds by arguing to a common principle (as opposed to a cause) behind the phenomena.

But before giving my account of classical inductive inference, I will make a suggestion about the nature of explanation. It counts as a virtue in an explanation that many phenomena are referred to a *single* cause or other principle.

But is this virtue an additional or an essential virtue? My tentative suggestion is that it is actually of the essence of a good explanation that it unifies phenomena by referring to a common principle. On any view, this is a common form of explanation. My suggestion is that it is involved in all explanation.

A very common form of explanation is to explain how something *works*: to give the mechanism by which the something produces whatever it does. But, one might wonder, does giving a mechanism involve giving a unifying principle? It can be argued that it does. To point to the mechanism is to point to something working according to principles which it is assumed are known. Given these principles, together with the boundary conditions, that is, the set-up of the mechanism, the mechanism works as it does. The mechanism is thus shown to fall under known principles. But this is unification. The new is assimilated to an already given principle.

4. THE NATURE OF INDUCTIVE INFERENCE

The way that inductive inference is to go should now be fairly clear. The inductive evidence may be that all observed *F*s are *G*s, where *F* and *G* are dummies for suitable predicates. (More on their suitability later.) This is a molecular state of affairs, a conjunction of "atomic" state of affairs involving the individual *F*s. This coincidence of the properties *F* and *G* in all observed instances is exceedingly improbable *if* there is no further factor in the situation which determines an *F* to be a *G*. But, if, to take the simplest case there is a deterministic *law* linking something's being an *F* with that thing being a *G* then the probability that *any F* is a *G* can be set at 1. This hypothesis of a nomic connection of properties can then be tested by observing further *F*s.

The law that we are talking about is of course not the law-*statement*, "It is a law that *F*s are *G*s," but is that state of affairs in the world which makes that statement true. It is essential that the law be some "atomic" state of affairs, because otherwise it will not be a *single* principle whose existence explains the molecular state of affairs that each observed *F* is a *G*. If however the law is, ontologically, nothing but the universal conjunction of *F* and *G*, dear old $(\forall x)(Fx \supset Gx)$, then the single-principle pattern of inference is unavailable. It is no explanation of some *F*s being *G* that *all F*s are *G*. If all *F*s are *G*s then it can be deduced that all observed *F*s are *G*s, but explanation is more than just deduction. So if the law is nothing but a universal conjunction, then inductive inference to unobserved cases cannot proceed via the law. I doubt that it can proceed at all, but arguing for that negative is not my business here.

Notice, and this is a point about which it is easy to be confused, that the proposed account of the inductive inference is *not* any account which *presupposes* that a Humean or a regularity account of the nature of laws is mistaken, and that instead the universe is governed by strong laws. Rather, the argument goes *from* the observed constant conjunction of characteristics *to* the existence of a strong law, and thence to a testable prediction that the conjunction will extend to all cases. The first inference is, of course, not valid one. Furthermore, it is logically possible that it should yield a true conclusion (all unobserved *F*s are *G*s) although there are in fact no strong laws. But that situation is no different from any other abductive scientific inference.

It is true, nevertheless, that there is much to be said against a Regularity or "cosmic coincidence" account of what a law is. In my 1983 book I assembled no less than sixteen criticisms of what I called the *Naive* regularity theory of laws. For the most part the criticisms are unoriginal. Yet the sixteen are, to adapt the poet Robert Browning a little, "sixteen damnations each sure if the other fails." There can be more sophisticated versions of the Regularity theory which perhaps meet or by-pass a few of these objections. But the rest remain, and the sophisticated versions are open each to additional, if varying, reproaches.

Of course, difficulties have been raised for strong theories of law also, and it would be necessary to answer these for my account of inductive inference to be complete. I cannot do this here. But I will just say this. The fact that the postulation of strong laws leads to such a simple and natural pattern of inference from the observed to the unobserved via the law is itself a (metatheoretical) reason for thinking that the notion of a strong law is both intelligible and free from self-contradiction.

Premature self-gratulations, however, is to be avoided. Not any version of a strong law of nature will suffice as the *tertium quid* postulated to mediate between the observed and the unobserved. Some upholders of strong laws have suggested that the form of a law may be no more than this:

$$\boxed{\mathrm{N}}(\forall x)(Fx \supset Gx)$$

Here the operator $\boxed{\mathrm{N}}$ stands for some kind of necessity, nomic, causal or even absolute, stronger than mere universality.

This, however, is quite unsuitable for our purposes. It would be satisfied by a multitude of suitable necessities holding in each individual instantiation (positive instantiation) of the law ($\boxed{\mathrm{N}}(Fa \supset Ga)$, $\boxed{\mathrm{N}}(Fb \supset Gb)$ ...). But if that is the case, then the state of affairs appealed to as lying behind the observed uniformity will be as molecular as the state of affairs to be explained. It will not have the unity, the "atomic" nature, that is required for it to function as a *common* "cause".

I do not myself see any way to achieve the required unity except to conceive of the law as a relation holding between properties, properties which are universals. Taking the second point first, the properties must be universals: no substitutes can be accepted. For instance, equivalence classes of exactly resembling *particularized* properties ("tropes") serve as a rather good substitute for universals in many contexts. But particularized properties will not serve here, because they do not yield the required unity. The law would become a series of connections holding between trope F_1 and trope G_1, trope F_2 and G_2 ... and so on.

There is no great hardship in having to postulate universals. Natural science talks all the time about properties and relations, many of which fall into quantitative ranges (the different masses, the different lengths, taken as types, not tokens). These properties and relations are rather naturally construed as universals, with the *same* mass or length being predicated of different particulars. If we are scientific realists we should favour an *a posteriori* realism about universals, postulating the existence of just those properties and relations required for whatever is the true scientific world-picture. The most worrying thing about universals is the intellectual pressure to postulate *uninstantiated* universals. I believe, however, that it is possible to get along without them. The most plausible argument known to me for such universals, due to Michael Tooley (1977, p. 669), seems to do no more, at best, than establish that uninstantiated universals are *possible*. But even that conclusion can be questioned (Armstrong 1983, chap. 8).

If we thus hold a scientifically oriented and, as David Lewis puts it, a *sparse* theory of universals (1983, p. 345), then we shall not expect that in every induction we reason from an observed regularity that is an instantiation of universals. Pre-theoretically, certain kinds strike us as natural kinds, certain properties as natural properties, and we infer on the basis of observations involving such kinds and properties. But we are prepared to learn that such epistemically aboriginal kinds or properties are no better than disjunctions of classes of universals. No doubt, these classes have a family nature: they are families of universals that are reasonable closely-knit, perhaps which are clusters of universals in the technical sense of "cluster" introduced by Douglas Gasking (1960). But, in general, the epistemically primitive kinds and properties are not universals.

We meet up with things that we classify as *F*s and to which we attribute property *G*. We are prepared to hear, therefore, that the actual nomic connection involves "hidden" universals F^* and G^* which precisify two *ranges* of universals represented by the "properties" F and G. Some classifications we will even thrust aside as not following the joints of the beast of reality. This shows us why the application of a predicate such as "grue" yields no foundation for inductive inference.

So the recognition of universals seems necessary for the proposed rational reconstruction of good inductive inference. But it is not sufficient. To get "atomic" states of affairs which explain the multiple instantiations of the law, we require a relationship between universals themselves. This relationship might take the form of something's being F necessitating or probabilifying that thing's being G. (Necessitation being thought of as probability strictly 1.) This relation automatically expresses itself in Fs being Gs, or in there being a certain probability of an F being a G. This relation between universals is the single factor (cause in a *very* wide sense) that lies behind the observed uniformity or frequency. The relation between the law and its (positive) instantiations may seem puzzling. I believe that the puzzlement is relieved if the law is seen not merely as a relationship between universals, but as itself a universal, fully instantiated in each positive instantiation of the law.

A question remains whether these relationships between properties that are universals are necessary or contingent connections. I favour the contingency view, but it seems that whichever view is taken, the account of the inference can remain the same. After all, even if, as I do not believe, laws are the sort of thing that hold "in every possible world", still the law can only be established *a posteriori* as a result of experience. As a result, inductive reasoning will still have to be used and there will still be a question what is the true structure of inductive reasoning. The account given in this paper will remain a candidate for the correct elucidation of that structure.

5. CONCLUSION

Has any real advance been made? In particular, has anything been done to help exorcize sceptical doubts about induction?

In the first place, it is surely a notable simplification to reduce induction to a particular species of abduction. Such a simplification is itself an echo of the Principle of the Common Cause, or rather the Principle of the Common Principle.

In the second place, induction conceived of as a mere transition from observed regularity to unobserved regularity, has always seemed a difficult and problematic inference. Why should an observed regularity be maintained? The

Principle of the Common Cause or Principle is by contrast a supremely natural principle, "comfortable to the human intellect", to lift a phrase from Keynes, who was talking about economics.

Induction conceived as abduction remains as invalid a form of inference as ever. But, as I have already explained, induction is so intertwined with the bedrock of our beliefs that there is no question of giving it up.

It is perhaps worth noticing that the particular account given of induction can be used against philosophers like van Fraasen (1980) who would have us give no credence to inferences beyond what is observable while still accepting them for practical purposes. One moral of my argument is that the principles of reasoning to the unobserved do not differ inside and outside the sphere of the observable. But if we trust these principles inside the sphere of the observable, why should we not trust the principles when they take us beyond that sphere?

Such are the advantages I see in this account of the nature of inductive inference. But there is what some will account a disadvantage, perhaps a crippling one. One will have to accept an inference that goes via strong laws, probably in the form of relationships between universals. For myself, of course, I think that we have here some argument for universals, and for strong laws that connect universals.

A direction for future research appears. We do not know very much about abductive argument, about inference to the best explanation. The notions are used a great deal in contemporary discussion, and their value is undoubted. But it is unclear what is the fine structure of such arguments. For instance, can they be, or to what extent can they be, formalized? I do not think that we know. It is an interesting prospect that research in this field may simultaneously advance the study of the Problem of Induction.

University of Sydney

NOTE

* This paper was first published in *Dialogue*, XXX 4 (1991), pp. 503–511.

REFERENCES

Armstrong, David M., 1983, *What is a Law of Nature?*. Cambridge: Cambridge University Press.

Buchler, Justus, ed., 1940, *The Philosophy of Peirce*. London: Kegan Paul.

Foster, John, 1983, "Induction, Explanation and Natural Necessity." In *Proceedings of the Ar-*

istotelian Society, N.S. Vol. 83, p. 87–101.

Gasking, Douglas A.T., 1960, "Clusters." *Australasian Journal of Philosophy*, 38: 1–36.

Lewis, David K., 1983, "New Work for a Theory of Universals." *Australasian Journal of Philosophy*, 61: 343–77.

Moore, G.E., 1925, "A Defence of Common Sense." In *Contemporary British Philosophy* (second series). Edited by J.H. Muirhead. London: Macmillan. Reprinted in G.E. Moore, *Philosophical Papers*. London: Allen & Unwin, 1970.

Reichenbach, Hans, 1956, *The Direction of Time*. Berkeley and Los Angeles: University of California Press.

Salmon, Wesley, 1978, "Why ask 'Why'?" In *Proceedings and Addresses of the American Philosophical Association*, Vol. 51, p. 683–705.

Smart, Jack C., 1982, "Difficulties for Realism in the Philosophy of Science." In *Logic, Methodology and Philosophy of Science VI: Proceedings of the Sixth international Congress of Logic*, Methodology and Philosophy of Science, Hanover 1979. Edited by L. Cohen, J. Los, H. Pfeiffer and K.-P. Podewski. Amsterdam: North Holland Publishing, p. 363–75.

Stove, David C., 1986, *The Rationality of Induction*. Oxford: Oxford University Press.

Tooley, Michael, 1977, "The Nature of Laws." *Canadian Journal of Philosophy*, 7, 4: 667–98.

van Fraasen, Bas, 1980, *The Scientific Image*. Oxford: Oxford University Press.

ALAIN BOYER

THE OSIANDERIAN COMPROMISE

As is well known, in 1543 Copernicus's *De Orbium Caelestium* was published with an anonymous foreword. This foreword, as Kepler clearly saw, was not Copernicus's, but had been written by the editor of the book, the Lutheran theologian Andreas Osiander.

In this short and lucidly written text, Osiander tries to convince the reader that Copernicus's hypotheses are not to be understood prima facie, as true or even thruthlike (*verisimiles*) descriptions of reality, but only as computation devices. Osiander's position is usually regarded as a clear expression of the so-called instrumentalist theory of science. As such, it has been criticized by realists like Alexander Koyré and Karl Popper, and on the other hand forcefully defended by modern instrumentalists like Pierre Duhem, in his famous short book, *Sôzein to Phaïnomena* (*To save the phenomena*).

I'd like to argue that Osiander's conception of the aim of Science belongs to a coherent ideological system, and that it can be interpreted as a tentative compromise between Science and Authority, academic and religious authorities. The crucial issue is truth, and the unity of truth.

* * *

Osiander's main motivation is clearly stated at the beginning of his text. He wants to defuse the bomb implicitly contained in the heliocentric, or more exactly heliostatic theory. Perhaps even better than Copernicus himself, he clearly sees that a realistic interpretation of the new astronomy can be regarded as threatening the coherence of the current world view:

> "Certain scholars, I have no doubt, are deeply offended and believe that the liberal arts, which were established long ago on a sound basis, should not be thrown into a confusion."

Regarded as a true or truthlike description of the world, the Copernican system could well endanger the whole system of the sciences, cancelling the current distinctions, and engaging in a risky reorganization of the relationships between astronomy, physics and theology. One can see that a question of the ontological interpretation of scientific theories might well turn out to be ideologically and even, I should add, politically relevant.

J. Misiek (ed.), The Problem of Rationality in Science and its Philosophy, 55–63.

* * *

In his *Idea of Nature*, Collingwood asserts rightly that it is a "philosophically foolish and historically false idea" to suppose that the philosophical significance of the new astronomy was to diminish the importance of man in the universe: "no philosophical problem (. . .) is at all affected by considering the relative amount of space (men) occupy" (p. 96)

But Collingwood goes on to say that its significance was rather "implicitly to deny that the world has a centre at all. As his posthumous editor (Osiander) said, you could regard any point as its centre; and for the purpose of studying the planetary orbits, it was convenient so to regard the sun. This statement has sometimes been regarded as due to timidity in the face of established doctrine, as if it amounted to saying "I admit that the orthodox view is true, but the heliocentric view is nevertheless a convenient fiction". But its real point was that the material world has no centre, and this was rightly regarded as a revolution in cosmology, because it destroyed the entire theory of the natural world as an organism".

This is a beautiful and ingenious interpretation. But I think it cannot be maintainded. Copernicus seems himself to have forcefully believed at least in the movement of the Earth, the crucial point, and in the central position of the sun; moreover, Osiander does not seem to imply in any sense that he wished Copernicus's work to be interpreted in such a bold way (the a-centricity of the world). I prefer to follow A. Koyré, who aptly characterized Osiander's foreword as "a little treatise of positivist and pragmatic epistemology".[1]

Collingwood seems to be nearer the truth when he claims that Copernicus's achievement was to produce a theory which requires a homogeneous conception of the physical world: "This idea, far from diminishing the scope of man's powers, vastly enlarged it; for it taught him that scientific laws established by him on earth would hold good throughout the starry heavens ".

P. Duhem, as is well know, provoked a certain amount of consternation when he argued that "we are compelled to recognize that Logic was on the side of Osiander, Bellarmine and Urbin the VIIIth, not of Kepler and Galileo". But he stressed that the (realist) Copernicans were right to require a unified physics: "The same dynamics must *represent* – with a unique set of mathematical formulae, the movements of the heavenly bodies, the waves of the Ocean, and the fall of the heavy bodies" (*op. cit.* Conclusion).

One could say that Duhem regards Galileo's and Newton's achievements as fulfilling J. Watkins's B_2 requirement of the Optimum Aim for Science, that is "to progress towards one unified science with theories that are increasingly unified".[2] But Duhem would have refused the B1 requirement, even, I submit, in its modified or "progressivist" version, that is, "to progress towards ever deeper explanation" (id.).

But, one could ask Duhem, is it plausible to imagine that an Osianderian strategy would have had enough heuristic power to provide, as the realist interpretation does, a powerful incentive towards a unifying Dynamics? It is usually said that the necessity of building a dynamics appropriate to a moving Earth was instumental in Galileo's research programm.[3] As against this, it seems to me that nothing in the Osianderian view requires any process of unification of any sorts. Nothing here implies that an instrumentalist terrestrial physics has to be compatible and, even less, organically connected with celestial physics, if only because each theoretical domain may require different assumptions. I conclude that the Osianderian strategy, as against Galileo's, would have less appeal for those who (like Duhem himself) are interested in the B_2 requirement. It is not sure, after all, that a unified physics would have to be simpler, easier to grasp or even more beautiful, than two juxtaposed ones.

According to Duhem, there is no doubt that the Osianderian point of view was not new: it looks like a restatement of one old Neo-Platonic position, expressed for instance by Proclus and Simplicius. For Duhem, Proclus in particular was a positivist of sorts. Nevertheless, one could argue with Popper that "Duhem claimed for instrumentalism a much older and much illustrious ancestry than is justified by the evidence". I would like to argue that Osiander's is a purer instrumentalist position than that of the Greek's, even if it remains a fictionalism: Osiander does not say that theories have no truth value, but that they are false and that it doesn't matter, because we do not demand them to be true or even truthlike. (Kepler in contrast argued strongly against the idea that false theories can explain true phenomena, and against the use made by anti - Copernicans of the rule "*ex falso verum sequitur*": like Aristotle, Kepler wants the explanans to be true, and known to be true.[5])

One could even show that Duhem mistranslates Proclus, when he gives "far from truthlike" for a greek expression which means something like "only plausible" or "only truthlike" (*eïkotos*).

The expression "*sozein ta phaïnomena*", in Simplicius, as well as "*apodosein ta phaïnomena*" in Aristotle, had as such no restrictive and positivistic connotations. It meant that irregular appearences have to be accounted for as the product of (possibly) true ordered movements. (I admit that Proclus' opinion seems to be more complex: but Proclus was a critic of Ptolemaic astronomy, from the point of view of truth, i.e., Plato.) The business of the astronomer does not reduce to thinking up computation devices but really consists in a demonstration of the fact that the heavens can be "saved" from the accusation of being imperfect. Astronomy is a kind of Theodicy (a polytheistic one, incidentally).

The Greeks had a strong fallibilist view of the human cognitive apparatus. They insisted that many different hypotheses are compatible with our finite evidence. Their view of science seems plainly hypothetico-deductivist. So, their *methodology* was not so alien to ours, except for the fact that they seem not to have really grasped the difference between *ad hoc* and independently testable hypotheses. The crucial question for them is: what constraints can realism – Physics or Metaphysics – exert on astronomical hypotheses ?[6] Indeed, the Greek metaphysical research program, which goes back at least as far as Plato, was peculiar to them. The minimal constraints exerted by realism on astronomy were, as is well known, at least twofold. Suppose an ontological diversity in the Universe, with:

1) the Earth at its bottom – the least dignified place – and at rest,
2) a perfect heaven, that is, an order governed by circular and uniform movement.

Every theory respecting these constraints and saving the phenomena (at least, accounting for observed facts without doing violence to them, as Aristotle says,[7] is a good candidate for truth.

The Pythagorean models (Philolaos) and, even more, Aristarchus's one were rejected for not having respected the first requirement, and nobody, before Kepler, challenged the second. So there is may be no contradiction between Ptolemy's rejection of Aristarchus in the name of realism and his unashamed acceptance of an instrumentalist interpretation of the choice of astronomical hypotheses, constrained only by the cosmologically true principle of circular uniform movement. The epistemological debate in medieval science was mainly concerned with the question whether these two constraints were sufficient and whether Ptolemy's astronomy enough conformed to the second one. Osiander's subtle strategy consists in using the astronomer's traditional responses to these metaphysical injonctions to show that the defence of free hypothesizing can be extended even to the position of the Earth, thus violating the first principle. One fixed point in realism had to disappear. It is not without interest to note that in his Address to the reader, he never refers to the second principle – the circular uniform movement, however forcefully this is sustained by Copernicus himself. One could perhaps imagine such a (non-Keplerian) foreword prefacing Kepler's *Astronomia Nova*. After all, the ellipse made the job of the astronomers much more simpler and was even able to save some phenomena better. From an Osianderian, but not from a strict Copernican point of view, Kepler's theory is much better. As Koestler and Lakatos-Zahar argue, Copernicus was a dogmatic Greek thinker of sorts. (In particular, he wanted to dispense with equants.)

So, Osiander's view is a much purer expression of instrumentalism than Simplicius' and Proclus' were. These were more or less concerned with truth. As men are fallible, in Xenophanes' sense, we can only hope for approximation: "*to engus*", Proclus himself said, with no hope of ever *proving* any astromical theory.

* * *

Osiander proposed a kind of compromise with the theologians – and hence with the religious authorities – and with the philosophers as well: Science has nothing to do with truth, ontology, and causality. We leave these questions – the most important ones – to you, so that you need not be afraid of any danger from us. Science will not put itself foreward as an arbiter on any profound issue. It cannot possibly threaten the ideological coherence of the current world view. There is such a sharp demarcation between science and metaphysics that no competition between them is possible. The criterion of demarcation is equivalent to a distinction between different aims.

Remember that Popper says that concerning their relative claims to be true, he sees no difference between science and metaphysics, and that the problem of truth is far more important than the problem of demarcation.[8] For him, the demarcation, which is not sharp, has to do with the method – the means – which is used, rather than with the end which is pursued. Some of the traditional aims of metaphysics, if modified in a fallibilist spirit, can be maintained.

As against this view, Positivists hold that the demarcation is a radical partition between two watertight domains. What constitutes positivism properly speaking seems to me to be the agreement on this watertightness. Perhaps less important is the question whether the domain of metaphysics contains pure nonsense, deeply significant truths, or deeply significant nonsense, as Popper describes Wittgenstein's position.[9] Positivism in this sense being the result of a tentative compromise, it is not strange to find, amoung those attached to it, people of very different orientations: apologists for Religion – from Bellarmine and Berkeley to Duhem, as well as sceptical agnostics or staunch atheists (most of the modern positivists).

The Osianderian compromise was a subltle one, because in giving truth to the other camp, as it were, it made it possible to make science independent of metaphysics, and create a completely autonomous domain, with its own criterions of goodness – don't say truth – and its own rules of intersubjective deliberation.

Osiander's tract has to be understood as a plea for free thinking in a completely autonomous scientific community. We do not claim to speak the truth: hence we can be our own judges. This move consists in exchanging, as it were, the claim to truth for freedom of thought. The demarcation strategy would result in the constitution of a closed but free community, independent of all external principles and emancipated from any tutelary authority, detached from history and living its own life.

* * *

In a nutshell, I'd like to propose the idea that what Galileo refused is precisely this Osianderian "historic compromise", explicitely proposed to him by Bellarmine (himself perhaps not a committed instrumentalist). This refusal exposed Galileo to the charge that heliocentrism is contradicted by Biblical evidence, that is, Truth. Of course, the Church could accept the idea of the necessary *interpretation* of the Holy Book when it is too crudely contradicted by empirical evidence. Biblical texts can be submitted to a hermeneutical reading, by which one distinguishes a manifest sense, as it were, valuable for "the vulgar", speaking its language (that of imagination), and a latent one, which alone is really true. But, *first*, this work has to be done by churchmen, who must carefully and not publically discuss the matter, and, secondly, one has to *prove* that the Bible needs to be re-interpreted. As you know, Galileo was unable to *prove* the movement of the Earth, even with his (false) theory of the tides, and the fact is that he didn't hesitate to interpret the Bible by himself. So, the real problem was political rather than metaphysical: who has the right to speak the truth on the real nature of the Universe, the mathematician or the theologian? And, who has the right to impose new interpretations of the Bible? Duhem regretted Galileo's refusal of the Osianderian compromise and it is true that it was hopeless to try to *prove* Copernicus's theory. But we can consider this refusal to be an important step in the modern overthrow of the whole ancient hierarchy of sciences and truth-speakers. To refuse the Osianderian compromise means to refuse the abandonment of deep truth and of explanation as aims of Science. It means that on many issues Science is bound to contradict the old system of thought.

* * *

I don't think this implies that Science must necessarly destroy on even replace faith (that would be Scientism.) On the contrary, I believe that what is required by modern science is what I have called, referring to Popper, a secularized epistemology.[10]. I mean that epistemology must keep off all authoritarian concepts, distinguish clearly belief and knowledge, and renounce the quest for certainty. One has to recognize that this quest is not compatible with the quest for deep truth. It is impossible to maximize both certainty and truth-content, because this would be tantamount to committing the fallacy of double maximization, when two components of an aim, as J. Watkins says, "sometimes or always pull in opposite directions, so that a move that is progressive relative to one component is retrogressive relative to the other".[11].

A secularized science, or, if you prefer, a radically humanistic science, does not presuppose the possibility of Omniscience, that is, of God.[12] Moreover, the modern scientific outlook needs no extraneous hypothesis, if only because such hypotheses are too likely to be ad hoc (One could quote Laplace's and Darwins's achievements as counting as successes in (tentatively) explaining facts for which one previously appealed to God's intervention: "Sire, je n'ai pas besoin de cette hypothèse".) Hence, Science is methodologically atheist, even if it is ontologically agnostic. We have to investigate Nature's nature – and History's nature as well – *as if* it had no meaning, *as if* it was a self-contained and blind system of laws and events. The compromise, if any, is not between a world of phenomena and a world of things – in – themsleves, but between and inquiry into the truth about the whole Universe and, possibly, a completely purified religious faith, not at all concerned with the "Secrets of Nature", as Pascal' said, but only with what we *ought* to do: the autonomy of Galilean Science relative to any moral world view implies its neutrality and, by the same token, the autonomy of ethics.

* * *

Duhem saw himself that a completely instrumentalist view cannot easily fit the phenomenon of the discovery of "novel facts". For instance, he relates the beautiful story[13] of the unintended discovery of "Poisson's spot" after Poisson's derivation of its existence from Fresnel's theory. Duhem speaks of the "clairvoyant divination" of the physical theory, indicating that it is not only a classification of empirical laws, but a "natural" one, since "its arrangements lead us to suspect the real affinities between things".[14] I agree with A. Musgrave that there is here at least "a whiff of realism".[15] I conclude that instrumentalism, or anti-realism, is likely not to be a stable position: whereas its positive requirements are perhaps the weakest one can imagine, it has the flow of conducing to make quasi-realist assumptions when confronted with the problem of explaining the heuristic power of theories without having recourse to chance, or to miracle.

* * *

My conclusion will focus on the problem of the philosophical "neutrality" of methodologies. J. Watkins (*op. cit.*, p. 124) regards as a requirement for any proposed aim of science that of *impartiality*. The strong point of this requirement is the idea that "a proposed aim for science (. . .) should not favour one side (in the scientific battle) because of a bias towards theories that endorse a particular metaphysical view of the world" (*id.* p. 125). We sould not "make Kant's mistake of elevating certain striking characteristics of a reigning theory into preconditions for any future theory". One could perhaps say that the Optimum Aim for Science has to be discussed under a quasi-rawlsian "veil of ignorance" concerning the content of current scientific theories, including the metaphysics imbedded in them. (I'd like to add that Kant's mistake is may be not only a Koenigsbergian idiosyncrasy, but that it is, so to speak, a natural tendency of the philosophical understanding.) Anyway, the problem is, does not the ideal of *depth*, advocated by John Watkins, go against the neutrality requirement? Is a discussion of the aims of Science possibly neutral in the debate between realism and anti-realism ?

The preference for depth has been historically defended by realists and the "anti-depth war", in John Watkins'words, always was linked with anti- realism. So, even if a sophisticated instrumentalist can argue that he favours "deep" theories and not only observable terms, I would call "methodological realism" (M. R.) the search for depth. M. R. does not imply a realist interpretation of theories, and does not *imply* that science can tell the truth about any "hidden" reality, no more than methodological determinism implies ontological determinism.[16] But it seems to me that, were this realist interpretation utterly misconceived, the undubitable achievements of M. R. would be something like a pure historical fact, leaved unexplained.

Popper states that realism is not a presupposition of his methodology, but that "it forms a kind of background that gives point to our search for truth, (which) would be pointless without an objective reality".[17] Nature has no sense, but Science has one. Why not *hope* at least that we can not only Save the phenomena, but, doing that, try to save the world? We don't have to commit the fallacy of misplaced economy, so to speak. As the French philosopher (of Polish origin) Meyerson said, criticizing Comte and Duhem, "the ontological character of scientific explanation is indelible".[18] Moreover, metaphysical realism seems to be a natural premiss for explaining the success of science, and then our ability to "use" it in practice. If one wants to save Science from scepticism, one has to combine, so to speak, "Realism and the Aim of Science".

Université Blaise Pascal, Clermont-Ferrand
and CREA, Paris

NOTES

[1] *La Révolution astronomique*, Hermann, Paris, 1961, p. 39.
[2] *Science and Scepticism*, Princeton University Press, 1984, p. 130.
[3] Cf. I. B. Cohen, *The Birth of a New Physics*, Norton, New York and London, 1985, p. 81.
[4] *Conjectures and Refutations*, R.K.P., London, 1963, p. 99, n. 6.
[5] Cf. Koyré, *op. cit.*, II, I, p. 130 sqq.
[6] Duhem, *op. cit.*, Vrin, Paris, 1990, p. 17.
[7] Cf. *De Caelo*, 293a25, quoted by Popper (*loc. cit*).
[8] Cf. *Realism and the Aim of Science*, Hutchinson, London, 1982, p. XIX and Part. I, ch. II.
[9] Cf. *The Open Society*, R.K.P., London, 1962, ch. 11, n. 51.
[10] Cf. Alain Boyer, *K. Popper: une épistémologie laïque*, Presses de l'Ecole Normale Supérieure, Paris, 1978.
[11] *Op. cit.* p. 124.
[12] Cf. K. Popper, "Indeterminism in Quantum Physics and in Classical Physics", *B.J.P.S.*, 1950, p. 38.
[13] A story used by J. Watkins to illustrate the Popperian notion of a "severe test".
[14] *La Théorie physique*, Vrin, Paris, 1981, ch. II
[15] Cf. his "Realism versus Constructive Empiricism", in *Images of Science*, ed. P. Churchland and C. A. Hooker.
[16] Cf. K. Popper, *The Open Universe*, Hutchinson, 1982, p. 149, with a reference to David Miller.
[17] *R.A.S.*, p. 81.
[18] *Identité et Réalité*, Vrin, Paris, 1910, p. 440; a quotation used by Quine.

ADELINO CATTANI

POPPER, POLANYI AND THE NOTION OF RATIONALITY

POLANYI'S ENTHYMEMATIC NOTION OF RATIONALITY

Polanyi's interpretation of rationality, far from being relativistic or irrationalistic, is realistic and usable. I would call it an 'enthymematic concept of rationality', founded on some 'rhetorical' notions, such as tacit premisses, consensus, plausibility, cooperation, conviviality, all of which indicate a mutual consent and a common agreement, reached by means of persuasion, based on convergent opinions not irrevocable and beliefs not final (commitment/objectivism, rhetoric/science, rhetoric/logic are the pairs and the framework in which the discussion can be inserted).

If rhetoric recovers its ancient rational dimension, we may say that Polanyi brings again the unarticulated, unexpressed elements in the quasi-logical dimension, replacing critical philosophy with a 'post-critical' one. He believes that only authority and trust (that is, acquired knowledge and principle of sincerity) guarantees a true communication[1] and that speaking and writing is an ever renewed struggle to be both apposite and intelligible (that is, using Paul Grice's terminology, relevant and perspicuous).[2]

POLANYI AND POPPER

Polanyi has little in common with Popper. They share aversion for every kind of dogmatism, ideological or epistemological, and advocacy of a liberal society. But the Popperian 'open society' does not correspond to the 'free society' of Polanyi. The latter is a society 'in which men, being engaged in various activities whose ends are considered worthy of respect, are allowed the freedom to pursue these ends. A free society is therefore one whose citizens in the main are committed to – dedicated to – various ideal ends (such as truth) and therefore one that is able to respect the free activities of its citizens in pursuit of such ends. It cannot be a free society by being open on such matters as these, that is, by being neutral with respect to truth and falsehood, justice and injustice, honesty and fraud'.[3] That is possible because the authority exercised in a society of explorers is mutually imposed and rests on mutual control, in

J. Misiek (ed.), The Problem of Rationality in Science and its Philosophy, 65–74.

science as well in politics, in art as well in ethics. The idea of a mutual cross-control promoting truth in science and progress in society is an influential suggestion first made by Polanyi. Truth would be not a property of a belief, but a result of a confrontation leading to consensus between members of a community.

Both Popper and Polanyi agree that research is neither an inductive process from observations, nor a mere deductive one from theory; but the starting point for Popper is a problem, while for Polanyi it is tacit knowledge (unarticulated or inarticulable): 'nothing is a problem or a discovery in itself; it can be a problem only if it puzzles and worries somebody, and a discovery only if it relieves somebody from the burden of a problem'.[4] So as a clever chess player finds unproblematic a solution puzzling for an amateur, the proofs supporting a theory change according to the individuals involved, their expectations, background knowledge, bias, beliefs, etc.

He polemically points out Popper's teaching which advocates a dispassionate science: an idea clearly expressed by its idealization of falsification, which imposes upon the scientist not only an attitude of indifference towards his own suppositions, but one which should also induce him to search for self-refutation. But the equation of science with refutability has many counter-examples.

> This is not only contrary to experience; it is logical nonsense as well. The surmises of a working scientist are born of his imagination in the quest for a genuine discovery. Such an effort risks defeat by submitting these surmises to rigorous tests, but never seeks it. . . Courts of law employ two separate lawyers to argue opposite pleas, because it is only by passionate commitment to a particular view that the imagination can discover the evidence that support it.[5]

The supposition that a scientist will choose the least probable hypothesis to investigate, since it will be the easiest one to refute, that is, the most risky, 'borders on the ridiculous. . . The cart has not only been placed before the horse, but the horse has been turned completely around'.[6]

Popper's view is criticized also because it discredits all ethical statements: 'No conceivable occurrence, no measurement or observation, can decide whether any action is moral or immoral, just or unjust, good or evil. . . As long as science remains the ideal of knowledge, and detachment the ideal of science, ethics cannot be secured from complete destruction by skeptical doubt'.[7] If you say that it is wrong to bear false witness, you say something that cannot be proved or disproved – according to Popper's amendment – by experiential facts.

We should say that in Polanyi's opinion, Popper's idea of rationality is unrealistic, impracticable, too demanding and restrictive. If Popper is the philosopher of perspicuity (the explicit and the criticizable), for whom the ineffable and the unexpressed are extra-logical (psychological, sociological) matters, Polanyi wants to bring the unstated within logic, or at least within human rationality, where criticism is important but not subject to fixed criteria of acceptability: the tacit component, in science too, is not a dispensable sediment, but an important epistemological ground. This means that all our saying presupposes and implies, all that is unsaid by us.

If rationality is transparency and if the peculiarity of science is the suppression of the implicit, then the unexpressed must be eliminated from rational-scientific activity. But if the tacit foundations are conditions for, not simply roots of, our knowledge, the Cartesian-Popperian view must be revised. The anti-Cartesian thesis that we know more than we can tell (tacit vs. explicit, subsidiary vs. focal awareness) means that our speech springs from an act (Polanyi) and flows into an act (Austin), whose appreciation is not possible in terms of 'criticism'. Systematic forms of criticism can apply only to articulated and expressed forms.

Polanyi recognizes the value of the critical movement in the past four or five centuries, but today is seems near the end of its course; now we must go back to St. Augustine, toward a post-critical philosophy or a fiduciary programme, grounded on the *nisi credideritis*, *non intelligetis*: 'We must recognize belief once more as the source of all knowledge. Tacit assent and intellectual passions, the sharing of an idiom and of a cultural heritage, affiliation to a like-minded community: such are the impulses which shape our vision of the nature of things on which we rely for our mastery of things. No intelligence, however critical or original, can operate outside such a fiduciary framework'.[8]

POLANYI CRITICIZED

In spite of his frank philosophical interests (see the opening sentence of his *Tacit Dimension*), Michael Polanyi is not included in philosophical texts. (In Italy, it is worth noting, his name is unfamiliar and the polyhedric Michael, a practicing scientist turned philosopher, is often identified with the economist Karl Polanyi.) Usually he is associated with Hanson, Toulmin, Kuhn and Feyerabend, the team who are critical of Popper's idea of scientific rationality and, with them, he is charged with irrationalism.

Rationality is generally contrasted with want of reason (arbitrariness, unreasonableness), lack of criticism (injudicious and fantastic thought), dogmatism (blind authoritarianism) or with pre/post/trans/anti-rationalism. In what sense would Polanyi's position be irrationalist? Has Polanyi come down on the side of anti-rationality? Why has his position been charged with obscurantism and catholic dogmatism? Is Polanyi arguing from ignorance? And is tacit knowledge inevitably tacit?

To associate Polanyi and Kuhn, whose best-known works appeared the same year (1962), is not unjustified.[9] Correspondingly Kuhn acknowledges that the basic assumptions of Polanyi are very similar to his own:[10] he asserts that the commitment to a paradigm of normal science does not requires the expression of rules and assumptions, a thesis brilliantly developed in chapters V and VI of *Personal Knowledge*, where it is argued that the scientist's success depends on 'tacit knowledge', that is on knowledge acquired in practice and impossibile to articulate explicitly. Finally, both link the problem of evaluating scientific theories (internal intellectual history) according to the reasons the scientists, members of a community, have for accepting an authority or a paradigm (external social history). An 'objectivist' framework cannot account for the appreciation (acceptance or rejection) of a scientific theory.[11]

Kuhn and Polanyi are criticized as supporters of an unsound, conservative authoritarianism. Such a charge came from Agassi, who finds Polanyi's theory authoritarian because 'philosophically opposed to any rule; he thinks that scientific method is ineffable and its practice is a matter of the intuitions of leading scientists';[12] Perelman too, who agrees with Polanyi, is afraid of the élitist and counter-revolutionary nature of his theory; in Lakatos's opinion, Polanyi's proposal is intellectually and politically conservative.

Admitting no general theory of scientific rationality, but only some local rational appraisals, 'Polanyi is a conservative rationalist concerning science, and an 'irrationalist' concerning the philosophy of science. But, of course, this meta-'irrationalism' is a perfectly respectable brand of rationalism: to claim that the concept of 'scientifically acceptable' cannot be further defined, but only transmitted by the channels of 'personal knowledge', does not make one an outright irrationalist, only an outright conservative. Polanyi's position in the philosophy of natural science corresponds closely to Oakeshott's ultra-conservative philosophy of political science'.[13] There is a great deal of truth in denying any universal and explicit criterion of scientific rationality, demarcation and progress: no statute law (normative and aprioristic) at all, only case law (the verdict of a jury constituted by the scientists themselves) exist for deciding what is scientific (a criticism for which Lakatos has high regard, in spite of its élitism). A scientist can have a 'tacit foreknowledge of yet undiscovered things. This is indeed the kind of foreknowledge the Copernicans must have meant to affirm when they maintained passionately, against heavy pressure, during one hundred and forty years before Newton proved the point, that the heliocentric theory was not merely a convenient way of computing tha paths of planets, but was really true'.[14]

If a large part of scientific, and also methodological, knowledge involves a 'tacit dimension', 'one can do science but there is no point in philosophizing about it' and 'the layman cannot be a judge in appraising scientific theories: the tacit dimension is shared and understood only by the élite. Only they can judge their own work'.[15]

Polanyi seems not to have grasped the critical function of language (methodological debate and critical discussion of scientific standards), as Popper did. In Lakatos's view 'there must be a dual system of authority, because the wisdom of a scientific jury has not been, and cannot be, fully articulated by the philosopher's law. Laws need authoritative interpreters'.[16]

Lakatos contends against Polanyi especially for his conception of the lawless closed autocracy of science.[17]

These accusations ignore the apparent contradiction of a science imposing authoritative pronouncements (accepted views, current values) and, at the same time, encouraging creative dissent (novelties, originality).[18]

W.T. Scott, in a paper devoted to the Polanyian view of science,[19] asserts that 'Tacit knowing is an important part of logical reasoning' and 'does by any means deny the significance of logical argument or provide a way to dispense with it. . . Furthermore, the premisses of an argument and the meaning of the results are often interpreted intuitively'. I would say and translate: 'interpreted by means of rhetoric'.

Polanyi's theory is appealed to by proponents of the ideal of 'truth by (changing) consensus', of a truth resulting from communal opinions and reached by dialogical agreement. The ideal of a truth based on consensus is a concept of a rhetorical kind and Polanyi's work is apt to reconcile experimental with rhetorical conceptions of science, as, for example, urged by B. Brummet in his 'Postmodern Rhetoric'.[20]

I would also contend, supported by Polanyi's tenets, that an imperfect knowledge is also fully rational: often we believe and we know something we don't perfectly understand. Polanyi's tenets suggest the possibility of using some rhetorical concepts for interpreting scientific rationality and the behavior of scientists, theory and practice in science. This interpretation is grounded on the assumption that scientific language spring from ordinary language and on the belief that we can know more than we can prove.

The thesis that what is not said is part of meaning, as well as what is said, is symmetric with the thesis that what is not said is received and understood as what is said. The point then is to understand how it so happens. And whether in science the case is the same or not.

In Polanyi's opinion, 'science functions like art, religion, morality, law and other constituents of culture'.[21] This assertion is criticized by Chaïm Perelman, otherwise sympathetic with Polanyi, in his paper on 'Polanyi's interpretation of scientific inquiry'.[22] Both advocate personal knowledge, both see the importance of culture and of tradition, in opposition to contemporary positivism, but Perelman is less inclined than Polanyi to assimilate the personal commitment of the scientist to other kinds of commitment and he wants to save the peculiarity of sciences in our culture.

What is acceptable as part of science and why should we prefer science to astrology? These old questions (echoing Feyerabend's 'What is so special about science?') are answered by Polanyi in a interesting way, neither dogmatically nor trivially.

'An affirmation . . .will be the more valuable to science, the more it possesses: (1) certainty (accuracy); (2) systematic relevance (profundity); (3) intrinsic interest. The first two of these criteria are inherently scientific, the third is extrascientific. The three criteria apply jointly, so that deficiency in one is compensated by excellence in others'[23] And 'I do not entertain explanations in term of astrology, for I do not believe them to be true'.[24]

These statements would permit us to using some explanatory concepts Grice introduces in *Logic and Conversation*: the problem of justification is not purely a logical one, but pragmatic, conversational, communicational. That is particularly evident in the case of unexpressed, tacit, suppressed or enthymematic premises. Polanyi transfers to scientific quarters too the awareness that is absolutely normal, to leave in the background and unsaid many things, and to imply/implicate some data, some standards, values, procedures, rules (e.g. *onus probandi*), etc.

The ineliminable presence of the implicit make every inference hazardous (the holistic thesis of Duhem-Quine was a first token of that) and forbids us to entertain the idea of a perfect rationality. The best we can propose is an 'enthymematic rationality', in the double sense of 'enthymeme' (not fully expressed reasoning – an inference with unexpressed premisses or conclusion – or reasoning which is only probable), that is a 'rhetorical syllogism', according the aristotelian definitions. Polanyi's epistemology is not, then, an irrationalistic or mystical one, but rather a proposal of a rhetorical-hermeneutical rationality (where rhetoric and hermeneutic are two sides of the same coin).

ENTHYMEMATIC – POLANYIAN ARGUMENTATION IN SCIENCE

According to the traditional definition, an enthymeme is a syllogism incompletely stated, in which the premises or the conclusion are tacitly present but not expressed. The argument 'Socrates is mortal because he is a man' silently assumes the additional premise 'All men are mortal' and this addition makes the reasoning a perfect deductive argument. Any kind of argument can be expressed enthymematically and almost every real (vs. formal) argument involves tacit premises assumed, but not actually stated. Usually in actual discourse premises are omitted because they are accepted or acceptable (obvious, taken for granted) to the audience and then it would be unnecessary to state them. Sometimes it a precautionary measure. Often some items of the argument remain unexpressed for lack of clearness. Sometimes because we cannot state them.

In his *Introduction to Logic*, Irving M. Copi[25] indicates the importance to scientific progress of formulating explicitly all relevant assumptions in any hypothesis. That light travels in straight lines was assumed in the arguments of Columbus and Copernicus about the spherical shape of the Earth, starting from what appears to a watcher on the land (the decks of a receding ship will disappear from view before the masthead), but it was a hidden assumption. It is important to drag hidden assumptions into the open, because there is no chance to examine them critically and to decide whether they are true or false, or better, simply acceptable, if, using Peirce's famous phrase, we call truth the opinion which is is fated to be agreed upon by all who investigate. Progress is often achieved by formulating explicitly an assumption which had previously been hidden and then scrutinizing and rejecting it. So a great amount of background information, counting as premises or reasons for a claim, is 'enthymematically' involved in observational-experimental science too, as in reasoning, argumentation, inference.

Aristotle calls 'enthymeme' a rhetorical syllogism. That means: a shortened one or a merely probable one.

In the first sense, an enthymeme would be an argument one of whose elements has been left unstated and that can become a categorical syllogism by adding the premise or the conclusion omitted. A long reasoning is hard to follow, in Aristotle terms, and then, for purposes of persuasion, we should not mention all premises, if they are familiar: 'we must not carry its reasoning too far back, or the length of our argument will cause obscurity: nor must we put in all the steps that lead to our conclusion, or we shall waste words in saying what is manifest'.[26] He gives the following example: to show that Dorieus has been victor in a contest for which the prize is a crown, it is enough to say 'For he has been victor in the Olympic games', without adding 'And in the Olympic games the prize is a crown', a fact which everybody knows.[27]

In the other, most fundamental, sense, an enthymeme is a syllogism based on probable premises, having a persuasive rather than a demonstrative function.

Kuhn's paradigm can partially be identified with Polanyi's tacit knowledge that scientists possess, a knowledge suggestive rather than determinate, illuminative rather than explicit. This kind of knowledge embedded in paradigms is important for the continued pursuit of science and for understanding scientific practice and theory, even if this idea may seem objectionable.

Polanyi's point, that we can know far more than we can ever tell, rests on some empirical phenomena: we are able to recognize physiognomic features without knowing how we do it; we can acquire some conditioned responses without being aware of the stimulus; the empathetic process is indispensable for understanding the behaviour of other individuals; we recognize a word by tacit knowledge, as is shown by the fact that a word loses its meaning if we attend repeatedly to the pronunciation of the term. More generally, the necessity of tacit knowledge results from the argument that we could not fully formalize or objectivize our knowledge of reality: 'in order that we may formalize the relations that constitute a comprehensive entity. . . this entity. . . must be first identified informally by tacit knowing'.[28]

Another instance of tacit knowledge is our knowledge of scientific problems (and of its solutions) as such. How do we perceive a puzzle as such, recognize a problem as such? No theory of knowlege can say how that happens, because our knowledge is richer than all our theories of knowledge.

Even the sciences cannot do without a large component of nonconceptualized ('rhetorical') knowledge. Harvey's discovery of the circulation of the blood is an instance of Polanyi's thesis that a scientist's capacity to pursue valid knowledge of a problem, guided by his sense of approaching the solution, that a valid anticipation of the yet indetermined implications of the discovery arrived at the end[29] are accounted for by the tacit knowledge. It is still an object of debate for historians and epistemologists how Harvey reached his discovery, how he justified it and how/why this 'unnecessary and ridiculous', 'paradoxical, absurd, impossible, incomprehensible, injurious to human health' (as it was defined by Jean Riolan and Guy Patin, eminent french physiologists) was soon accepted. After centuries of studies on Harvey, his case is always problematic, because the genesis, the justification and the acceptance of the circulation of the blood follow ways not compatible with the standard view of scientific procedure. Harvey's proof of the circulation was an 'enthymematic' argumentation, based on an intuitive judgement, rather than a demonstration, a conceptual invention rather than a discovery: the 'visual' demonstration will be offered some decades after the publication of the *Exercitatio Anatomica de Motu Cordis et Sanguinis* (1628) by Malpighi (1661) and Van Leeuvenhock (1688), while Harvey could not show the connection between arteries and veins and miscalculated the quantity of blood circulating in the human body (calculation on which his theory rests): all this make more relevant Harvey's achievement and success, in spite of the disbelief of several authoritative opponents and on the ground of an argumentative enthymematic certainty instead of physiological evidence. The movement of the blood in a circle is a product of Harvey's imagination, stimulated by an Aristotelian suggestion as to atmospheric circularity.

Those who feel the threat of irrationality, justify the appearance in science of certain 'fortuitous' aspects, which cannot be logically or conceptually construed, such as suggestions, analogies, metaphors, argumentation, ingeniousness, skills, bias, insights, by confining them to the 'context of discovery' (how an idea come to a scientist), leaving to the 'context of justification' (how he proves or verifies/falsifies the statements that he makes) real, grounded, reliable scientific knowledge. Such demarcation seems not apt to explain scientific practice. In the light of what Polanyi has said about personal knowledge and the tacit dimension and in the light of the historical inspection, we must admit that a discovery becomes meaningful for a scientist only on the basis of his experience. An enlargment of the meaning of the concept of 'rationality' requires us to comprise both contexts, what is tacit and what is explicit, the conceptualized and the nonconceptualized kinds of knowledge, the subjective craft and the formulated rules. That is why the epistemologies of perspicuity can explain neither what nor how we know. Polanyi's tacit/explicit couple is not an incompatible alternative: the first term only has application if the second does and *vice versa*. Nor it is an exhaustive opposition. (The couple has been compared with the polar concepts used by other thinkers such as Wittgenstein (form of life/theory), Ryle (knowing how/knowing that), Chomsky (competence/performance).)

The cognitive-deductive approach to argument reconstruction is unsuccessful because the problem of identifying tacit assumptions, missing premises and presuppositions is not purely a logical one. There may be different ways to fill the gap, following different criteria and principles (charity, fidelity, nonredundancy, plausibility, etc.). A Polanyian approach to the problem of supplying missing premises and presuppositions, a problem far from being settled, may be another useful, non-purely epistemic way, toward a more practical and informal model of scientific and ordinary rationality.

Università di Padova
Istituto di Storia della Filosofia

NOTES

[1] M. Polanyi, *Personal Knowledge* (*PK*), London 1958, p. 206.
[2] M. Polanyi, *PK*, p. 207.
[3] M. Polanyi and H. Prosch, *Meaning*, Chicago 1975, pp. 196–97.
[4] M. Polanyi, *PK*, p. 122.
[5] M. Polanyi, *The Tacit Dimension* (*TD*), ch. 3 and *Meaning*, p. 195.
[6] M. Polanyi, *Meaning*, p. 225.
[7] M. Polanyi, *Meaning*, p. 27.
[8] M. Polanyi, *PK*, p. 266.
[9] *The Structure of Scientific Revolutions* (*SSR*) is quoted by Polanyi (*Meaning*, pp. 56–57) as supporting and confirming his own position, elaborated from the ideas of Poincaré, Hadamard, Polya and surprisingly anticipated in the nineteenth-century by William Whewell.
[10] Th. Kuhn, *SRS*, ch. 5, note 1.
[11] M. Polanyi, *PK*, pp. 16–17.
[12] J. Agassi, *Science in Flux*, Dordrecht 1975, p. 130.
[13] I. Lakatos, 'History of science and its rational reconstructions', in *Philosophical Papers*, vol. 1, Cambridge 1978, p. 130, n. 3.
[14] M. Polanyi, *TD*, 1966, p. 23.
[15] I. Lakatos, *Philosophical Papers*, vol. 2, p. 111 and 227.
[16] I. Lakatos, *ivi*, p. 153.
[17] I. Lakatos, *ivi*, p. 124, n. 3.
[18] M. Polanyi, *Meaning*, p. 188.
[19] W.T. Scott, in *The World and I*, August 1987, pp. 619–21.
[20] B. Brummet, 'Postmodern Rhetoric', *Philosophy and Rhetoric*, 9, 1, 1976.
[21] M. Polanyi, *PK*, p. 133.
[22] Chaïm Perelman, *Le champ de l'argumentation*, Bruxelles 1970, pp. 344–52.
[23] M. Polanyi, *PK*, pp. 135–36.
[24] M. Polanyi, *PK*, p. 150.
[25] Irving M. Copi, *Introduction to Logic*, New York 1953, p. 425.
[26] Aristotle, *Rhetoric* 1395 b 25.
[27] Aristotle, *Rhetoric* 1357 a 16.
[28] M. Polanyi, *TD*, p. 21.
[29] M. Polanyi, *TD*, p. 24.

PAUL M. CHURCHLAND

ON THE NATURE OF EXPLANATION: A PDP APPROACH*

1. INTRODUCTION

The notion of explanation has figured centrally in most contemporary accounts of scientific knowledge and rational belief. Explanation is usually cited, along with prediction, as one of the two principal functions of our factual beliefs. And the rationality of such beliefs is commonly said to be measured, at least in part, by the relative range or quality of the explanations they make possible. If something like this is correct, then it is important for us to try to understand what explanation is, and what distinguishes a good explanation from a poor one.

Several existing accounts attempt to meet this challenge. They will be addressed below. The present paper proposes a new account of the matter – the *prototype activation model* – an account distinguished, for starters, by its being grounded in a novel and unorthodox conception of what cognition consists in. That conception derives from current research in cognitive neurobiology, and from Parallel Distributed Processing (PDP) models of brain function. (See Rumelhart *et al*. 1986a, 1986b; Churchland 1986, 1989.) These PDP models are noteworthy for many reasons, but first among them in the present context is their almost complete dissociation from the *sentential* or *propositional* conception of what knowledge consists in, and from the *rule-governed inference* conception of human information processing. Those venerable conceptions play a central role in all of the older accounts of explanation, and in orthodox accounts of cognition generally. They will play almost no role in the account to be proposed.

This is not to say that the older accounts of explanation are entirely without integrity. On the contrary, from the PDP perspective we can see why each of them has the appeal it does, and also why that appeal is to some degree genuine. That is to say, the prototype activation model successfully reduces/explains the major features of its competitors. It is also free from most of their celebrated defects. I will try to illustrate these claims as we proceed.

The prototype activation model is focused first and foremost on what it is to have *explanatory understanding* of a problematic thing, event, or state of affairs. The linguistic expression, exchange, or production of such understanding, should there be any, is an entirely secondary matter. We shall approach the

J. Misiek (Ed.), The Problem of Rationality in Science and its Philosophy, 75–108.

Kluwer Academic Publishers.

topic with the aims of an empirical scientist rather than with the aims of a logician or conceptual analyst. The goal is to outline a substantive empirical theory of what explanatory understanding really is, rather than to provide an analysis of the concept of explanation as it currently is or ideally should be used. Normative issues will certainly be addressed, and some unexpected insights will emerge, but normative concerns are not the prime focus of the present paper. Rather, what concerns us is the nature of the cognitive process that takes place inside the brain of the creature for whom explanatory understanding suddenly dawns, and in whom it is occasionally reactivated.

On the prototype activation (PA) model, a close approximation to this process is the process of *perceptual recognition*, as when one suddenly recognizes an indistinct outline as the face of a close friend, or as when one finally recognizes the faint motion under the hedge as a foraging mouse. On the PA model, essentially the same kind of computational achievement underlies both perceptual recognition and explanatory understanding. The latter is distinguished primarily by being a response to a wider variety of cognitive situations: it is not limited to sensory inputs.

A close connection between perception and explanation is by now a familiar theme in both psychology and philosophy. One's perceptual judgments, and perhaps even one's perceptual experiences themselves, have often been portrayed as the perceiver's best *explanatory* account of the peripheral stimuli (Gregory 1966, 1970; Rock 1983). In this tradition the notion of explanation is used in hopes of explicating the phenomenon of perception. The strategy of the present paper will reverse the order of things somewhat: we shall exploit a novel PDP account of perceptual recognition in hopes of explicating the phenomenon of explanatory understanding. I remain faithful to the earlier tradition, however, since my basic aim will be to show that both phenomena are fundamentally the same.

Let me open the discussion by trying to motivate the search for a new account of explanation, and for trying to launch it in the specific directions indicated. We may begin by recalling the 'Covering-Law' or 'Deductive-Nomological' (D-N) model (Hempel 1965), since so many of the current models are just artful restrictions or generalizations of that basic and very elegant idea. While much attention has been paid to the *logical* virtues and vices of this model, relatively little has been paid to its shortcomings when evaluated from a *psychological* point of view.

In fact, the D-N model is psychologically unrealistic in several important ways. If someone has just come to understand why *a* is *F*, the D-N model typically requires that we ascribe to that person knowledge of some universally quantified conditional statement having *Fx* as its consequent, plus knowledge of a series of initial conditions adequate to discharge the conjuncts in the antecedent of that conditional, plus the successful deduction of *Fa* from this assembled information, or at least the appreciation that the deductive relation exists.

However, while people have an explanatory understanding of much of what goes on around them on a minute-by-minute and even a second-by-second basis, people are decidedly and regularly inarticulate when asked to voice either the general law on which their understanding is presumably based, or the set of initial conditions that tie that law to the explanandum then at issue. What premises people are typically able to supply, when queried, often falls dramatically short of the full requirements of the D-N model. Moreover, the logical acumen ascribed to people on this account is often substantially in excess of what university students with formal training in logic can display.

Furthermore, the assembly of relevant factual premises and the search for relevant deductive relations is a process that takes time, often a good deal of time. All of this sits poorly with the great speed with which explanatory understanding is commonly achieved. It is often achieved almost instantaneously, as when one understands at a glance why one end of the kitchen is filled with smoke: the toast is burning! Such swiftness is not confined to mundane cases. If one has the relevant conceptual skills, the same speed is also displayed in more esoteric cases, as when one appreciates at a glance why Jupiter is an oblate spheroid: it is a plastic object spinning rapidly; or as when one appreciates at a glance why some red giant close-binary star has the shape of an egg pointed at its more compact blue companion: it is a very large object free-falling in a gravitational field.

At the other end of the spectrum, nonhuman animals provide a further illustration of these difficulties. Animals too display behavior that indicates the achievement of explanatory understanding, as when a frustrated coyote bites and paws at the leg-trap whose jaws have captured its mate. The coyote understands why its mate cannot leave. Animals too can anticipate elements of the future and understand elements of the present and past, often in some detail. But the assembly of discursive premises and the execution of formal inferences is presumably beyond their capacities, especially at the speeds that faithfulness to their insight and behavior requires.

These particular criticisms of the D-N model are unusual in being empirical and psychological, rather than logical in character. Even so, they are highly general. They will apply to all of the accounts of explanation that require, as the original D-N model requires, extensive propositional knowledge, relevant retrieval of same, and keen deductive insight. For it is precisely these features that give rise to the difficulties. Is there some alternative way of characterizing the way knowledge is represented in living creatures, and the way it is deployed or accessed in specific cases to provide local explanatory understanding? Yes, there is.

2. CONCEPTUAL ORGANIZATION IN PDP NETWORKS

Recent years have seen an upswell of research into the functional properties of 'neural' networks. These are artificial networks that simulate or model certain salient features of the neuronal organization of the brain. What is interesting is that even simple versions of these networks have shown themselves capable of some very striking computational achievements, and they perform these computations in a fashion that recalls important features of animal cognition.

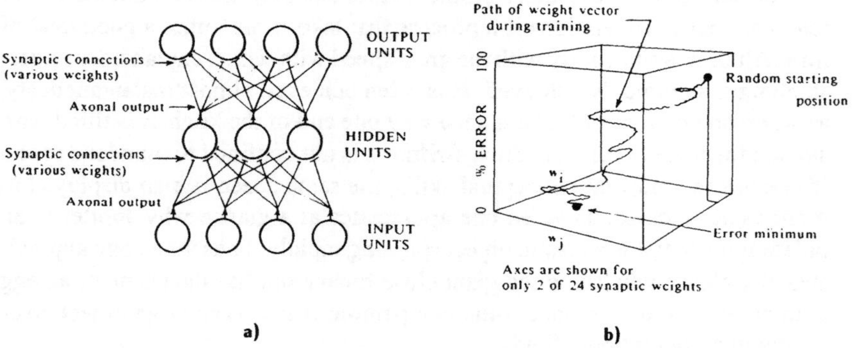

Figure 1. a) A simple feedforward network. b) Learning: gradient descent in weight/error space.

A primary feature of neuronal organization is abstractly depicted in the 'neural network' of Figure 1a. The circles in the bottom row of the network represent a population of sensory neurons. Each of these units projects a proprietary axonal fiber toward a second population of neuron-like units. The axon there divides into a fan of terminal branches, so as to make a synaptic connection with every unit in the second population. Real brains are not quite so exhaustive in their connectivity, but a typical axon can make many thousands or even hundreds of thousands of connections.

This arrangement allows any unit at the input layer to have an impact on the activation levels of all of the units at the second or 'hidden' layer. An input stimulus produces some activation level in a given input unit, which then conveys a signal of proportional strength along its axon and out the end branches to the many synaptic connections onto the hidden units. These connections stimulate/ inhibit the hidden units, as a function of the strength of the signal, the size or 'weight' of each synaptic connection, and its polarity. A given hidden unit simply sums the effects incident from its many input synapses. The global effect is that a *pattern of activations* across the set of input units produces a distinct *pattern of activations* across the set of hidden units. Which pattern gets produced, for a given input, is strictly determined by the configuration of synaptic weights meeting the hidden units.

The units in the second layer project in turn to a third population of units, there to make another set of synaptic connections. In real brains this pattern is typically iterated through many layers (roughly, $5 < n < 50$) before the chain concludes in a population of motor or other 'output' neurons, but for purposes of illustration a network of just three layers will suffice.

In this upper half also, the global effect is that an activation pattern across the hidden units produces a distinct activation pattern across the output units. As before, exactly what pattern-to-pattern transformation takes place is fixed by the configuration of synaptic weights meeting the output units. All told, this network is a device for transforming any one of a great many possible input vectors (i.e., activation patterns) into a uniquely corresponding output vector. It is a device for computing a specific function, and which function it computes is fixed by the global configuration of its synaptic weights.

Now for the payoff. There are various procedures for adjusting the weights so as to yield a network that computes almost any function – that is, any vector-to-vector transformation – that we might desire. In fact, *we can even impose on it a function we are unable to specify*, so long as we can supply a modestly large set of examples of the desired input/output pairs. This process is called 'training up the network'. It typically proceeds by entering a sample input vector at the lowest layer, noting the erroneous vector this produces at the topmost layer, calculating the difference between this actual output and the desired output, and then feeding this error measure into a special rule, the generalized delta rule (Rumelhart *et al.*. 1986a). That rule then makes a small adjustment in the antecedent configuration of all of the synaptic weights in the network. Repeating this procedure many times, over the many examples, forces the network to slide down an error gradient in the abstract space that represents its possible synaptic weights (see Figure 1b). The adjustments continue until the network has finally assumed a configuration of weights that does yield the appropriate outputs for all of the inputs in the training set.

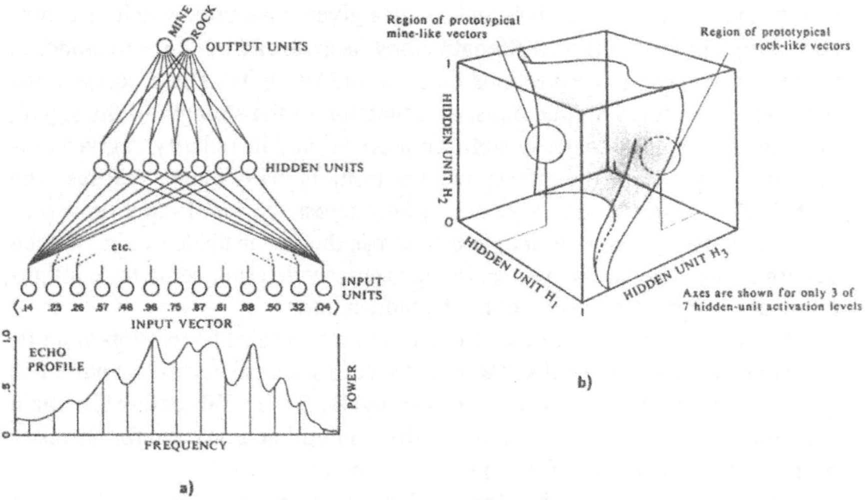

Figure 2. a) Perceptual recognition with a large network. b) Learned partition on hidden unit activation-vector space.

To give a real example, suppose we want the network to discriminate sonar echoes returned from large metallic objects, such as explosive mines, from sonar echoes returned from large submarine rocks. The discrimination of such echoes poses a serious problem because they are effectively indistinguishable by the human ear, and they vary widely in character even within each class. We begin by recording fifty different mine echoes and fifty different rock echoes, a fair sample of each. We then digitize each echo with a frequency analyzer, and feed the resulting vector into the bank of input units (Figure 2a). We want the output units to respond with appropriate activation levels (specifically, { 1, 0 } for a mine; { 0, 1 } for a rock) when fed an echo of either kind.

The network's initial verdicts are chaotic, since its synaptic weights were set at random values. But under the pressure of the weight-nudging algorithm, it gradually learns to make the desired distinction. And after it has mastered the echoes in the training set, it will generalize: it will reliably identify mine and rock echoes from outside its training set, echoes it has never heard before. Mine echoes, it turns out, are indeed united by some subtle weave of features, to which weave the network has become tuned during the training process. The same is true for rock echoes (see Gorman and Sejnowski 1988).

Here we have a binary discrimination between a pair of diffuse and very hard-to-define acoustic properties. Indeed, *we never did define them!* It is the network that has generated an appropriate internal characterization of each type of sound, fueled only by examples. If we now examine the behavior of the hidden units during discriminatory acts in the trained network, we discover that the training process has partitioned the space of possible activation vectors across the hidden units. (See Figure 2b. Note that this space is not the space of Figure 1b. Figure 1b depicts the space of possible synaptic weights. Figure 2b depicts the space of possible activation vectors across the middle layer.) The training process has generated a *similarity gradient* that culminates in two 'hot spots' – two rough regions that represent the range of hidden-unit vector codings for a *prototypical* mine and a *prototypical* rock. The job of the top half of the network is then just the relatively simple one of discriminating these two subvolumes of that vector space.

Several features of such networks beg emphasis. First, the output verdict for any input is produced very quickly, for the computation occurs in parallel. The global computation at each layer of units is distributed among many simultaneously active processing elements: the weighted synapses and the summative cell bodies. Hence the expression, "parallel distributed processing". Most strikingly, the speed of processing is entirely independent of both the number of units involved and the complexity of the function executed. Speed is determined solely by the number of distinct *layers* in the network. This makes for very swift processing indeed. In a living brain, where a typical information processing pathway has something between five and fifty layers, and each pass through that hierarchy takes something between ten and twenty milliseconds per layer, we are looking at overall processing times, even for complex recognitional problems, of between 1/20th of a second and one second. Empirically, this is the right range for living creatures.

Second, such networks are functionally persistent. They degrade gracefully under the scattered destruction of synapses or units. Since each synapse supplies such a small part of any computation, its demise leaves the network essentially unchanged. Third, and very important for our purposes, the network will regularly render correct verdicts given only a degraded version or a smallish part of a familiar input vector. This is because the degraded or partial vector is relevantly *similar* to a prototypical input, and the internal coding strategy generated in the course of training is exquisitely sensitive to such similarities among possible inputs.

And exactly which similarities are those? They are whichever similarities meet the joint condition that (a) they unite some significant portion of the examples in the training set, and (b) the network managed to become tuned to them in the course of training. The point is that there are often many overlapping dimensions of similarity being individually monitored by the trained network: individually they may be modest in their effects, but if several are detected together their impact can be decisive. Here we may recall Wittgenstein's description (1953) of how humans can learn, by ostension, to detect 'family resemblances' that defy easy definition. PDP networks recreate exactly this phenomenon.

Since that early period, various theorists have independently found reason to introduce similar notions in a number of cognitive fields. They have been called "paradigms' and "exemplars" in the philosophy of science (Kuhn 1962), "stereotypes" in semantics (Putnam 1975), "frames" (Minsky 1981) and "scripts" (Schank and Abelson 1977) in AI research, and finally "prototypes" in psychology (Rosch 1981) and linguistics (Lakoff 1987). These vague but persistent ideas appear to find microphysical explication, and systematic vindication, in the partitions and focused similarity gradients produced across the activation-vector spaces of trained networks.

Finally, such networks can learn functions far more complex than the one illustrated, and make discriminations far beyond the binary example portrayed. In the course of learning to pronounce English text, Rosenberg and Sejnowski's NETtalk (1987) partitioned its hidden-unit vector space into fully 79 subspaces, one for each of the 79 letter-to-phoneme transformations that characterize the phonetic significance of English spelling. Since there are 79 distinct phonemes in English speech, but only 26 letters in the alphabet, each letter clearly admits of several different phonetic interpretations, the correct one being determined by context. Despite this ambiguity, the network also learned to detect which of several possible transforms is the appropriate one, by being sensitive to the various letters that flank the target letter inside the word. All of this is a notoriously irregular matter for English spelling, but the 'rules' were learned by the network even so.

Other networks have learned to recognize the complex configuration and orientation of curved surfaces, given only gray-scale pictures of those surfaces as input. That is, they solve a version of the classic shape-from-shading problem in visual psychology. Still others learn to divine the grammatical elements of sentences fed as input, or to predict the molecular folding of proteins given amino acid sequences as input. These networks perform their surprising feats of learned categorization and perceptual recognition with only the smallest of 'neuronal' resources – usually much less than 10^3 units. This is only one hundred millionth of the resources available in the human brain. With such powerful cognitive effects being displayed in such modest artificial models, it is plausible that they represent a major insight into the functional significance of our own brain's microstructure. That, in any case, is the assumption on which the following discussion will proceed.

3. RECOGNITION AND UNDERSTANDING

Let me now try to highlight those functional features of PDP networks that will lead us back toward the topic of explanation. The first feature I want to emphasize is the partitioning, in a suitably trained network, of its hidden unit activation vector space into a system of prototype representations, one each for the general categories to which the network has been trained (see again Figure 2b for the simplest sort of case). Any prototype representation is in fact a specific vector (= pattern of activations) across the network's hidden units, but we may conceive of it more graphically as a specific point or small volume in an abstract state-space of possible activation vectors, since that portrayal highlights its geometrical relations with representations of distinct prototypes, and with activation vectors that are variously close to (= similar to) the prototype vector.

The second point to emphasize is that a single prototypical point or activation vector across the hidden units represents a wide range of quite different possible sensory activation patterns at the input layer: it represents the extended family of relevant (but individually perhaps non-necessary) features that collectively unite the relevant class of stimuli into a single kind. Any member of that diverse class of stimuli will activate the entire prototype vector at the hidden units. Also, any input-layer stimulus that is relevantly *similar* to the members of that class, in part or in whole, will activate a vector at the hidden units that is fairly close, in state-space, to the prototype vector.

In dynamical terms, the prototype position is called an *attractor*. We may think here of a wide-mouthed funnel that will draw a broad but delicately related range of cases into a single narrow path. This process is instanced in your ability to recognize a friend's face in any of a wide variety of expressions, positions, and conditions of viewing. Or in your ability to recognize a horse in almost any posture and from almost any perspective. These are exactly the sorts of capabilities displayed by suitably trained PDP networks.

A third point is to emphasize again that PDP networks are extraordinarily fast. Once trained, they achieve the 'recognitions' at issue in a matter of milliseconds. And they will make distinct recognitions, one after another, as fast as you can feed them appropriately distinct stimuli.

Turn now to the units at the output layer. In the stickfigure account of cognition I am trying to outline, these are to be conceived as driving or initiating some specific motor activity: perhaps something relatively simple, as in NETtalk, where the output vector codes a phoneme and actually produces, via a speech synthesizer, an audible sound. In a living creature, however, the output will typically be more complex, as when a dog's sudden olfactory recognition of a gopher initiates a routine of rooting and digging at the favored location; or as when a bird's sudden visual recognition of a stalking cat prompts it to initiate a sequence of wing motions that launch it into the air.

The picture I am trying to evoke, of the cognitive lives of simple creatures, ascribes to them an organized 'library' of internal representations of various prototypical perceptual situations, situations to which prototypical *behaviors* are the computed output of the well-trained network. The prototypical situations include feeding opportunities, grooming demands, territorial defense, predator avoidance, mating opportunities, offspring demands, and other similarly basic situations, to each of which a certain broad class of behaviors is appropriate. And within the various generic prototype representations at the appropriate level of hidden units, there will be subdivisions into more specific subprototypes whose activation prompts highly specific versions of the generic form of behavior. (Is the mouse eating sesame seeds? Or hickory nuts? Is it avoiding a cat? Or a hawk?) These various prototypes are both united and distinguished by their relative positions in the hidden-unit vector space. They are all close together, but they differ slightly in their positions along one or more of the relevant axes. These differences evoke relevantly different responses at the output layer.

As just outlined, this picture will inevitably recall memories of Behaviorism, for the perceptual environment is here portrayed as the fundamental control of motor behavior, and the link between the two will appear very stimulus-responsish to many eyes. But that construal of the basic character of the mechanisms at issue is oversimple and deeply misleading, as I shall try briefly to explain.

The simple networks of Figures 1a and 2a fail to portray some obvious augmentations that living brains clearly use. Real brains boast many successive layers of hidden units rather than just one – perhaps as many as a hundred layers along some pathways. Further, real brains divide into many distinct processing hierarchies working side by side on different problems. A brain is not a single network, but a committee of many cooperating networks – perhaps over a thousand of them in a typical mammalian brain. And most importantly for the present issue, the input to a given bank of hidden units comes not *just* from the sensory periphery, but from elsewhere in the brain itself. The brain is a *recurrent* network. The all-up input to any layer will almost always include some 'current context' information that derives from earlier processing elsewhere in the brain.

Unlike the models in the figures, almost any layer of units in the brain will receive some 'descending axonal projections' returning from the next layer of units upwards in its hierarchy, and it will often receive 'horizontal projections' from layers within the many adjacent networks working on related informational tasks. Accordingly, which prototype vectors get activated within a given layer is generally a function of diverse inputs, some of which arrive unvarnished from the sensory periphery, but many of which reflect concurrent activity elsewhere in the brain.

This frees the brain from the knee-jerk style of operation that worried us a few moments ago. Its ultimate behavior is a function of factors so many and so subtle, factors that interact in such highly volatile ways, that the brain's behavior has become predictable only in its broadest outlines and only for very short periods into the future. Moreover, the factors controlling behavior reside within the brain itself as much as in the external environment.

To return to the basic issue. We can now see how the brain can command a large and sophisticated repertoire of prototype activation vectors, each one representing some complex prototypical situation in the external world. We have seen how such vectors can be activated by the perceptual apprehension of even a smallish portion of the relevant external situation, and how those vectors can activate in turn behaviors appropriate to the entire external situation, and not to just the smallish part that was initially coded in perception.

I wish to suggest that those prototype vectors, when activated, constitute the creature's recognition and concurrent *understanding* of its objective situation, an understanding that is reflected in the creature's subsequent behavior. Of course, a creature may *fail* to recognize/understand its current perceptual or cognitive situation. The vector activated at the relevant layer of hidden units may fall well outside any of the prototypical volumes of the relevant state-space, and the behavior subsequently produced will therefore not be drawn from its well-honed repertoire. The resulting behavior may be just confused. Or it may be a default routine of flight from the unknown. Or perhaps it will be a default routine of stumbling *exploration*, one that may finally find either a physical or a cognitive perspective from which the situation suddenly does activate one of the creature's many prototype vectors. It may find, that is, a perspective from which the situation suddenly does make sense.

By way of whatever learning algorithm governs synaptic adjustments, such failures and subsequent successes, especially in quantity, will modify the character and state-space location of the creature's internal prototype representations, so that situations of the puzzling kind just solved will successfully activate a prototype vector more readily in future.

4. PROTOTYPE ACTIVATION: A UNIFIED THEORY OF EXPLANATION

The aim of the preceding sections was to illustrate the initial plausibility of a novel conception of cognitive activity, a conception in which vector coding and vector-to-vector transformation constitute the basic forms of representation and computation, rather than sentential structures and inferences made according to structure-sensitive rules. Let us assume, for the sake of argument, that this conception is basically accurate even for human brains. If so, then we must immediately be impressed by the range of conceptual resources such systems can command, given the neuronal machinery available.

With roughly 10^{11} non-sensory neurons, the human brain commands a global state-space of fully 10^{11} dimensions. Each brain subsystem will typically be operating with something like one-thousandth of that number, which gives a typical specialized state space approximately 10^8 proprietary dimensions to play with. This will allow for some stunningly complex and fine-grained representations, since a single vector with 10^8 elements can code the contents of an entire book. A state space of 10^8 dimensions will also allow for a similarly stunning variety of coding vectors. If we assume that each neuron admits of only ten distinct levels of activation (a serious underestimation), then that typical specialized state space must have at least 10^{10^8} or $10^{100,000,000}$ functionally distinct positions within it. This is the number of distinct possible *activation vectors*. To appreciate the magnitude of this number, recall that the total number of elementary particles in the entire physical universe, photons included, is only about 10^{87}. And recall that, on the above assumptions, your brain commands something like a thousand of these specialized state spaces.

Of course, the vectors themselves represent nothing, save in the context of the global configuration of synaptic weights that produced them, and dictate their effects on subsequent layers of neurons. Here again we have a superastronomical range of possibilities. In a given subsystem of 10^8 neurons, a typical neuron will have at least 10^3 synaptic connections arriving from other neurons, for a total of 10^{11} distinct synapses within that subsystem. If each synapse admits of only ten distinct weights, then we have $10^{10^{11}}$ or $10^{100,000,000,000}$ distinct possible configurations of *weights* for that subsystem alone.

Overall, and speaking crudely, this means that a typical subsystem will have a lexicon of 10^{10^8} possible representations, each of which has 10^8 elements, and each of which could have any one of $10^{10^{11}}$ possible 'meanings'. We should not balk, therefore, at the premise of the following discussion, which regards it as unproblematic that the brain should command intricate prototype representations of such things as stellar collapse, cell meiosis, positron-positron collision, redox reaction, gravitational lens, oceanic rift, harmonic oscillator, intentional action, and economic depression. Such phenomena, intricate though they are, are not beyond reach of the representational resources described.

The discussion to this point has all been preamble to the following sugges-

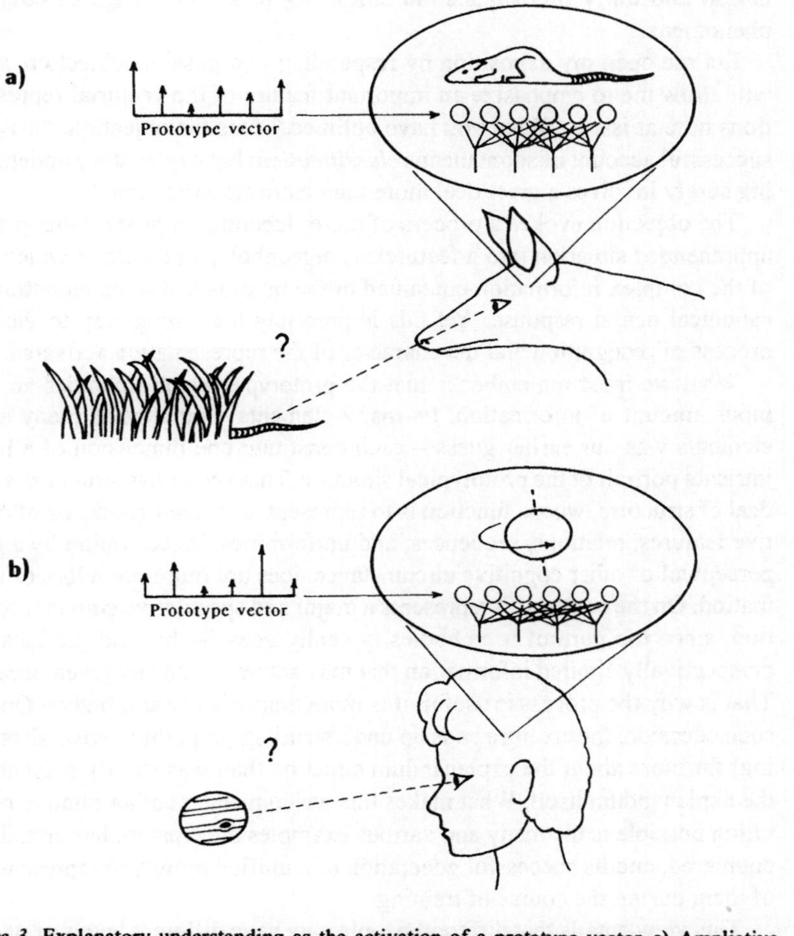

Figure 3. Explanatory understanding as the activation of a prototype vector. a) Ampliative activation of *desert rat* vector. b) Ampliative activation of *rotating plastic body* vector.

tion: explanatory understanding consists in the activation of a specific prototype vector in a welltrained network. It consists in the apprehension of the problematic case as an instance of a general type, *a type for which the experienced creature has a detailed and well-informed representation*. Such a representation allows the creature to anticipate aspects of the case so far unperceived, and to deploy practical techniques appropriate to the case at hand (see Figure 3 a & b). Given the preceding discussion, this idea has some plausibility already. It is my aim in the remainder of this essay to illustrate how much illumination and unity this suggestion can bring to a wide range of cognitive phenomena.

Let me open my exposition by responding to a possible objection, which will allow me to emphasize an important feature of the vectorial representations here at issue. "What you have outlined," runs the objection, "may be a successful account of spontaneous *classification*; but explanatory understanding surely involves a great deal more than mere classification."

The objection evokes a process of mere 'labeling', a process that puts the apprehended situation into a featureless pigeonhole, a process in which most of the complex information contained in the input is lost in its reduction to a canonical neural response. Yet this is precisely the wrong way to view the process of recognition and the character of the representation activated.

What we must remember is that the prototype vector embodies an enormous amount of information. Its many elements – perhaps as many as 10^8 elements was our earlier guess – each constitute one dimension of a highly intricate portrait of the prototypical situation. That vector has structure, a great deal of structure, whose function is to represent an overall *syndrome* of objective features, relations, sequences, and uniformities. Its activation by a given perceptual or other cognitive circumstance does not represent a loss of information. On the contrary, it represents a major and speculative *gain* in information, since the portrait it embodies typically goes far beyond the local and perspectivally-limited information that may activate it on any given occasion. That is why the process is useful: it is quite dramatically ampliative. On each such occasion, the creature ends up understanding (or perhaps *mis*understanding) far more about the explanandum situation than was strictly presented in the explanandum itself. What makes this welcome talent of ampliative recognition possible is the many and various examples the creature has already encountered, and its successful generation of a unified prototype representation of them during the course of training.

This view entails that different people may have different levels or degrees of explanatory understanding, even though they 'classify' a given situation in what is extensionally the 'same' way. The reason is that the richness of their respective prototype representations may differ substantially. This is a welcome consequence, since explanatory understanding does indeed come in degrees. On the present view, its measure is just the richness and accuracy of the creature's prototype.

With these points in hand, let us now turn to a larger issue. One prominent fact, ill-addressed by any existing account of explanation, is the variety of different *types* of explanation. We have causal explanations, functional explanations, moral explanations, derivational explanations, and so forth. Despite some procrustean analytical attempts, no one of these seems to be the basic type to which all of the others can be assimilated. On the prototype activation model, however, we can unify them all in the following way. Explanatory understanding is the same thing in all of these cases: what differs is the character of the prototype that is activated.

4.1. Property-Cluster Prototypes

I begin with what is presumably the simplest, most common, and most superficial kind of explanatory understanding, and with the simplest and most superficial kind of prototype: the cluster of typically co-occurrent properties. Think of the typical cluster of features that constitutes a cat, or a cookie, or a tree, or a bird. These prototypes comprehend the vast majority of one's conceptual population, and they are activated on a regular basis in the course of one's mundane affairs. Because of their familiarity to everyone, cases involving them are seldom puzzling to anyone. But the explicit questions of inquiring children reflect the background explanatory role these prototypes continue to play for all of us. "Why is its neck so long, Daddy?" "It's a *swan* dear; swans have very long necks." "Why is he all spotted, Mummy?" "He's a *leopard* dear; leopards are always spotted, except when they're young."

4.2. Etiological Prototypes

These are what lie behind *causal* explanations. An etiological prototype depicts a typical temporal *sequence* of event-types, such as the cooking of food upon exposure to heat, the deformation of a fragile object during impact with a tougher one, the escape of liquid from a tilted container, and so on and so on. These sequences contain prototypical elements in a prototypical order, and they make possible our explanatory understanding of the temporally extended world. We apparently command a very large repertoire of such prototypes, much of it organized into an elaborate hierarchy of species and sub-species.

Note that the temporal inverse of an etiological prototype is generally not an etiological prototype as well. This means that causal explanations are generally asymmetric in character. The height of a flagpole and the altitude of the Sun may jointly explain the length of the pole's shadow. But the length of the pole's shadow will serve to explain neither of the other two facts. That asymmetry, a major problem for other accounts of explanation, is a natural consequence of the present account.

Some may complain here that I have simply helped myself to an unexplicated notion of 'causal process' or 'etiology' in order to sustain a particular analysis of explanation, one that takes 'cause' as a primitive. This charge misses my purpose. I am not attempting to provide an analytic definition of the notion of explanation. I am trying to provide a revealing and unifying characterization of the sorts of brain events that constitute explanatory understanding. Those events, I suggest, are prototype activations. An important subclass of activated prototypes represent typical temporal sequences or processes. These, I observe, appear to underwrite what philosophers have called causal explanations.

Now just what intricacies constitute a genuine etiological prototype, and how the brain distinguishes between real causal processes and mere pseudo-processes, are secondary matters I shall leave for a future occasion. We have, at a minimum, a novel set of resources with which to address that old problem. It is sufficient for present purposes to observe that, within the framework of the PA model, we can immediately reconstruct, at least to a first approximation, the major types of antecedently recognized explanation. To this task I now return.

4.3. Practical Prototypes

These, I suggest, are what lie behind *functional* explanations. One thing humans understand very well, being agents in the world, is complex means/ends relations between possible situations realizable by us and expectable consequences thereof that may be desirable to us. To portray any temporal configuration of actual or potential situations in this means/ends way is to make graphic or salient for us certain of the causal relations that unite them. It is a way of portraying a causal structure in the world in the guise of an actual or figurative practical problem, with the explanandum event or feature as its practical solution.

Practical prototypes, like etiological prototypes, also depict sequences of event-types or feature dependencies, but in the case of practical prototypes the

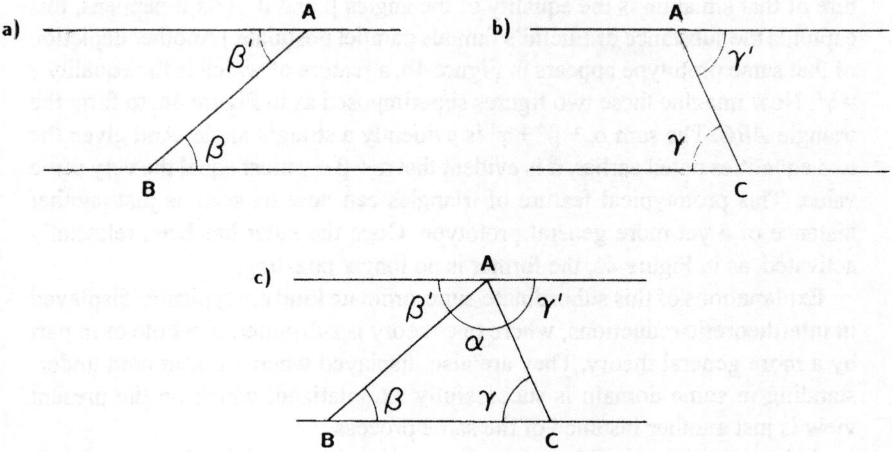

Figure. 4. Understanding a subordinate prototype as one instance of a superordinate prototype.

explanandum *begins* the explanatory sequence whereas in etiological prototypes the explanandum *concludes* the explanatory sequence. Thus, a functional explanation does provide some entirely objective information, as in "His peripheral blood vessels all contracted in order to protect his central organs from the gathering cold." But sometimes they ring very hollow, as in "Supernova 1987A collapsed on itself in order to spread photons and neutrinos throughout the physical universe." Just when functional explanations are appropriate and when they are not is a delicate issue I shall bypass on this occasion.

4.4. *Superordinate Prototypes*

Some explanations, typically those in science, logic, and mathematics, concern not singular facts but *general* truths. Why do planets move on ellipses? Why are the theorems of the sentential calculus all tautologies? Why do the interior angles of any triangle always sum to exactly one straight angle? Here the objects of puzzlement are not singular situations; they are prototypical syndromes themselves. Even so, explanatory understanding seems to consist in the apprehension of the subordinate prototype as being an instance of some superordinate prototype.

The puzzle about triangles will illustrate the point. Why is the sum always the same? And why is it 180° exactly? Consider the prototypical situation depicted in Figure 4a, viz., two parallel lines cut by a third. A prototypical feature of that situation is the equality of the angles β and β'. (As it happens, this captures the substance of Euclid's famous parallel postulate.) Another depiction of that same prototype appears in Figure 4b, a feature of which is the equality $\gamma = \gamma'$. Now imagine these two figures superimposed as in Figure 4c, to form the triangle *ABC*. The sum $\alpha + \beta' + \gamma'$ is evidently a straight angle. And given the two equalities noted earlier, it is evident that $\alpha + \beta + \gamma$ must equal the very same value. This prototypical feature of triangles can now be seen as just another instance of a yet more general prototype. Once the latter has been relevantly activated, as in Figure 4c, the former is no longer puzzling.

Explanations of this subordinate/superordinate kind are typically displayed in intertheoretic reductions, where one theory is subsumed in whole or in part by a more general theory. They are also displayed when our scattered understanding in some domain is successfully axiomatized, which on the present view is just another instance of the same process.

Axiomatizations, in this century's experience, are ruthlessly syntactic affairs. The D-N model was right at home in accounting for explanations within that general context. But axiomatization was not always thus. I have used an example from geometry here in order to suggest that the successful systematization of theoretical understanding can be pictorial or graphical, as well as occasionally syntactic. With many others, I acquired my conception of axiomatization as a student of formal logic and of the philosophy of science and mathematics in the mid-60's. Within that tradition, axiomatization was *essentially* a syntactic affair. When subsequently I looked through Euclid's original *Elements* – acclaimed as the first great axiomatization in history – I was therefore struck by how very *un*syntactic and how vividly diagrammatic the whole affair was.

This was standardly put down to the sloppiness of Euclid's presentation. I can no longer entirely believe that. Though quite real, syntactic systematization now appears as only one of several modes of cognitive systematization, perhaps of relatively minor importance. The basic mode of systematization is the reconfiguration of the similarity metric across the relevant vector space, so that previously diverse and unrelated cases are now re-represented as being relevantly similar to each other, as being slightly different instances of the same more general prototype. This reconfiguration can happen in two ways: within the space that contains the relevant prototypes, so that they come to occupy closely proximate positions; or in the next layer up, so that they are seen as instances of a higher prototype located in a subsequent layer of hidden units.

This more encompassing view of conceptual unification is of course entirely compatible with the idea of syntactic unification through axiomatization.

It merely subsumes that mode of organization under a more general mode. The presented 'axioms' of formal systems, you will recall, are almost always axiom *schemata*, having a potential infinity of distinct but recognizable instances. And the rules of inference are inference schemata as well, whose diverse applications must be mastered by the person using them. Thus do we learn to discriminate *proofs*. What unites the class of theorems is that each is the last line of some proof or other. In this way, evidently, can prototype-trained PDP networks handle the combinatorial and generative aspects of cognition. Their capacity for discriminating relevant structure and principled iteration may be limited, of course. But then, so is ours.

There is a great deal more to be said about explanatory unification, but for now let us move on.

4.5. Social-Interaction Prototypes

These underwrite *ethical*, *legal*, and *social-etiquette* explanations, a class of explanations unaddressed, to my knowledge, by any prior account of explanation. This is faintly surprising, since they are both real and familiar. "Why shouldn't I disperse this bunch of trouble-makers?" asks the red-neck cop approaching a so-far peaceful demonstration. "Because that would be a case of violating their constitutional right to peaceful assembly" is an explanatory reply. "Why", asks the gossip, "shouldn't I discuss Mary's marital problems with Doris?" "Because that would be violating a confidence" is an explanatory reply. "Aw, why can't I go play outside in the rain?", asks the seven-year-old boy. "Because you have your new clothes on, and company will be here for Thanksgiving dinner in half an hour", is an explanatory reply. Here we appeal to various prototypical misbehaviors – denying a constitutional right, betraying a confidence, being inappropriately turned out for a family fete – of which the contemplated behavior is claimed to be an instance. Acquiring a legal, moral, or social sensibility is a matter of mastering a large system of such social-interaction prototypes, and of coming to perceive the social world from within that framework. (Since writing this paper, I have learned that moral explanations have indeed been recognized and discussed. Sturgeon (1985), for example, assimilates them to the D-N pattern.)

4.6. Motivational Prototypes

These underwrite our familiar folk-psychological explanations of intentional actions. By rights, such explanations should not be assigned a separate category in the present list, since motivational prototypes are properly just a special subclass of the etiological prototypes discussed above in 4.2). They are, I believe, causal explanations (Churchland 1981a, 1988a). But I did not wish to break the flow of exposition at that point by addressing a disputed case. On the other hand, I do not wish to leave this important class entirely unmentioned.

Motivational prototypes depict typical configurations of desires, beliefs, and preferences as the antecedents of intelligent behavior. The explanations they underwrite are distinct from the functional explanations described in 4.3), since the elements in the causal sequence of a functional explanation are typically external to any agent's body, and temporally later than the action to be explained. With action-explanations, by contrast, the initial elements in the explanatory sequence are always internal to the agent – his desires, beliefs, and deliberative activities – and they are always prior to the behavior to be explained.

Motivational prototypes are but a subclass of a large family of affective, deliberative, and cognitive prototypes, those that collectively make up folk psychology. A perennial objection to the idea that our self conception has the character of a theory has been its apparent requirement that each of us command many thousands of universally quantified conditionals – the 'laws' of folk psychology. From the PDP perspective, this is no longer a requirement. As a picture is worth a thousand words, so a few hundred complex prototypes can do the work of many thousands of laws. And unlike laws, a prototype is activated directly by the apprehended situation it may serve to explain. The old 'access problem' largely evaporates. To be sure, inappropriate prototypes will often be activated, conflicts with further experience will thus ensue, and the puzzling explanandum will then have to be readdressed from within a more informed context. But the prototype vectors, whether appropriate or inappropriate, will typically be activated within a second or so.

The upshot for the dispute about Folk Psychology is that it is still a theory, just like any other theory. What has changed, and changed for the better, is our conception of how theoretical knowledge is both stored and accessed in cognitive creatures.

Though I very much doubt it is complete, I shall now bring this catalogue of prominent kinds of prototypes to a close. You can see how they allow the prototype activation model to account in a unified fashion for the most familiar and widely discussed types of explanatory understanding, and for some previously undiscussed types as well.

5. INFERENCE TO THE BEST EXPLANATION

The idea of prototype activation throws some much needed light on the popular idea of "inference to the best explanation", a process that has often been invoked to account for the fixation of many of our beliefs, even our observational beliefs (see, for example, Harman 1965, 1973). That idea is appealing, since it does seem to address what distinguishes the beliefs we do acquire from the many we might have acquired: the former have better explanatory power relative to the overall circumstance that occasioned them.

But the idea is also problematic, since it suggests a choice made from a range of considered alternatives. As a matter of psychological fact, alternatives are rarely even present. And in any case, our beliefs are typically fixed so swiftly that there is no time for the comparative evaluation of complex matters like the relative explanatory power of each of a range of alternatives.

On the PDP approach, we can begin to explicate the crude notion of "inference to the best explanation" with the more penetrating notion of "activation of the most appropriate prototype vector". Activating the most appropriate available prototype is what a well-trained network does as a matter of course, and it does it directly, in response to the input, without canvassing a single alternative vector. In the end, the process is not one of 'inference' at all, nor is its outcome generally a sentence. But the process is certainly real. It just needs to be reconceived within the more penetrating framework of cognitive neurodynamics. When it is, both the 'alternatives' problem and the 'speed' problem disappear.

C. S. Peirce, who called the process "abduction", found the former problem especially puzzling. Peirce, one of the pioneers of modern formal logic, appreciated very early that for any set of observations there is a literal infinity of possible hypotheses that might be posed in explanation. But how can we possibly search a space of infinite size? Indeed, how can we even *begin* to search it effectively when its elements are not well-ordered? Peirce marveled that human scientists are able so regularly to produce, from this infinite sea of mostly irrelevant and hopeless possible candidates, hypotheses that are both relevant and stand some non-trivial chance of being true. From the sentential perspective, Peirce was right to marvel. But from the neurocomputational perspective, the situation is not so mysterious.

We do not search an infinite space of possible explanations. In general, we do not search at all: in familiar cases a suitable prototype is activated directly. And if the novelty of the case foils our waiting categories, and thus forces us into search mode, then we search only the comparatively tiny space comprising the set of our own currently available prototype vectors. Even here the search is mostly blind, and probably stops at the first success. If one's initial encounter with the problematic explanandum fails to activate directly a familiar and subsequently successful prototype vector, then one repeatedly re-enters the problematic input in a variety of different cognitive contexts, in hopes of finally activating some prototype vector or other, or some vector close enough to an existing prototype to give one at least some handle on the problem.

Since the range of concurrently possible 'understandings' is closed under the relation "is at least within hailing distance of an existing prototype", then *of course* any element from that range will appear both 'relevant' and 'potentially true'. Peirce, and we, are the victims of a perspectival effect. Our hypotheses will look at least passably hopeful to us, because they are drawn from a source that collectively defines what will be found plausible by us. We should thus be wary of assuming, as Peirce seems to have assumed, that we have any special nose for truth.

I have raised Peirce's worry here for a further reason: the process just described illustrates the role of *analogy* in the discovery of new explanatory hypotheses. So many of our best theories have had their origins in a provocative initial analogy, perhaps strained in some dimensions, that subsequently proved to be a fruitful vehicle of understanding (see Hesse 1966, Kuhn 1962). The process finds a ready depiction within the present framework. A prototype vector whose activation has hitherto been confined to one empirical domain subsequently comes to be activated with profit in a new domain. More accurately, the new domain activates a vector that is very close to the old prototype, close enough to evoke at least some of its cognitive and behavioral consequences. An analogy between two domains has been discovered, and exploited.

Unfortunately, talk of 'analogy' has always been hobbled by our inability to say anything very specific about what constitutes the relevant kind of similarity. We are now in a position to be entirely specific. Analogy, or similarity, consists in the close proximity of the respective prototype representations, activated by the two empirical domains, in the relevant hidden-unit activation-vector space. And the constituting dimensions of that very high-dimensional similarity are given by the response properties of each neuron in the relevant hidden layer. This means that any analogy is a complex and highly subtle matter, easily recognized, perhaps, even though it will typically transcend effective verbal description. Most importantly, it is no longer a mystery why both of these things should be so. The many dimensions of the relevant vector space are *used* by us, but they are not generally *known* to us.

The title of this section refers to the "best" explanation. But how, on this PDP approach to explanation, are competing modes of understanding to be evaluated? What makes one explanation better than another? Here we must answer carefully, since we are denied the usual semantic vocabulary of reference, truth, consistency, entailment, and so forth. The cognitive kinematics here being explored does not have sentences or propositions as its basic elements; the basic elements are activation vectors. The various dimensions of epistemic *virtue*, therefore, will have to be reconceived in terms that are grounded in this new conception of what cognitive activity consists in.

That will be no small task, and I cannot pretend seriously to undertake it here. What I will do is illustrate how some aspects of the problem can be addressed, and how the network approach supplies unexpected insight into at least some perennial problems.

At the simplest level, it is easy enough to describe the differential virtue that two distinct prototype vectors *A* and *B* can have as a response to a problematic situation. If we assume that we are dealing with a well-trained network, and that the integrity of the two prototypes is not in question, then it may be a just criticism to say that *A* is simply the *wrong* prototype for the problematic situation at hand. It may be wrong because the situation confronted is not a member of the class of situations that will reliably activate *A* from almost any perspective, even though it happened to activate *A* on this occasion.

This can occur if the agent apprehended only a misleading part of the problematic situation, a part that led to the activation of *A* because that unusual part was relevantly similar to *A*'s typical activators. Unfortunately for the agent, however, the problematic situation is such that, if addressed from any one of many slightly different perceptual or cognitive points of view and then reentered into the network, it would reliably activate *B* in almost every case. In this statistical sense, *A misrepresents* the situation, whereas *B* does not.

Correlatively, the behavioral consequences typically activated in turn by *A* may be highly, even lethally, inappropriate to the problematic situation in question, in contrast to those activated by *B*. A real example of both failures would be a coyote's faulty understanding of the nature of a small tapered appendage disappearing into a tuft of long grass. The hungry coyote understands it as the tail of a retreating desert rat, and acts accordingly. In fact the problematic item is the tail of a poisonous snake, which coyotes generally avoid (see again Figure 3a).

Pragmatic and statistical considerations can thus provide a fairly robust distinction between appropriate and inappropriate prototypes, at least for simple cases. We can even reconstruct the distinction between the *correctness* of a prototype on a given occasion, and its *warrant* on that occasion. High warrant is a matter of low ambiguity in the input. We need to ask, is the input vector closely similar to any other possible input vector that would activate a different prototype? If so, then the ambiguity of the current input is high and the warrant of the prototype vector activated is correspondingly low. If the input is not remotely similar to any activationally diverse input, then its ambiguity is low and the warrant of its activated prototype is high. Similarity among input vectors is measured as usual: by their geometrical proximity within the activational state space for the relevant input layer.

The evaluation of occasion-specific explanations is one thing, but what of evaluating the propriety of an entire class of explanations? How do we evaluate, that is, the integrity of the prototypes, and systems of prototypes, themselves? It is difficult to provide much of an answer to this question, since the answer must draw on some general story of how to evaluate the global cognitive configurations of neural networks. It must draw, that is, on some analog of a general account of 'theoretical virtue'.

We are as yet in no position even to sketch such an account. But it is clear that there may be gold in these hills. The cognitive behavior of quite simple networks displays an unexpected connection between representational economy, conceptual unification, and the capacity for successfully generalizing past experience to novel cases. Very briefly, the story is this. Suppose we wish a network to discriminate *F*s. And suppose we train that network to a high level of success on some training set (the rock/mine network, for example). Despite its success at recognizing *F*s drawn from the training set, the network will do very poorly at recognizing new examples of *F* drawn from outside the training set, if the number of hidden units in the network is very much larger than a certain optimal number. The reason is that, with so many hidden units to exploit, the lazy network will tend to learn a distinct and unrelated prototype for each of the distinct and slightly different examples of *F* in its training set. These unrelated prototypes will sustain appropriate responses at the output layer, and with the error messages thus reduced to nothing, the system will cease any further learning. In particular, nothing forces it to generate a *single* prototype (at the center of a *unitary* similarity gradient) as its response to the entire range of relevantly similar examples of *F* in the training set.

The cost of such laziness comes due when the network is presented with a new example from the relevant class of stimuli, an example drawn from outside its training set. That example will be similar to all of the others, but the network is in a relatively poor position to see that similarity, since it has not learned to see *what is common* to all of the relevant examples in the original training set. To be sure, it will do better than chance at recognizing new examples, since it does have some scattered grasp of the class at issue. But still it performs poorly.

By contrast, a network that is denied a surplus of hidden units, a network given too few resources to permit such *ad hoc* and unprojectible learning, is a network that is forced to continue learning until it finds a single prototype region, at the center of a unified similarity gradient, a region near which to code every example of *F* in its training set. A *unified* solution to the recognition problem is the only solution for which it has adequate resources. After being forced by circumstance to find such a solution, the network will subsequently recognize a wide range of novel examples of *F* almost as reliably as those it was trained on, since it possesses a similarity gradient adequate to catch almost all of them. (For a more detailed discussion, see Churchland 1988b, or 1989a).

Conceptual unification, evidently, is a cognitive virtue of enormous importance, at least as conceived on the present model of cognition. It is important for the very good reason that cognitive configurations having that virtue do much better at generalizing their past experience to new cases. It is therefore reasonable to regard explanations that involve prototypes embedded in a unified cognitive configuration as superior to those that involve prototypes not so embedded. Here is a further dimension in which explanations can be evaluated.

All told then, we can evaluate them as follows. A virtuous mode of explanatory understanding (= an activated prototype vector) should be a *rich* portrait of the general type at issue; it should be strongly *warranted* (= low ambiguity in the input that occasions it); it should be *correct* (relative to the library of currently available alternative prototypes); and it should be part of the most *unified* cognitive configuration possible. Evidently, normative epistemology is not essentially tied to a propositional attitude conception of cognitive activity. It is robustly possible within the framework here being explored. And on the specific matter of conceptual unity – what it is, and why it is a virtue – the new framework can already claim a proprietary success, a success where the old framework has been a chronic failure.

6. COMPARISON WITH EARLIER MODELS

Let us begin with the venerable D-N model. It was correct in insisting that explanatory understanding requires the deployment of some information that is general in character. Beyond this insight, almost nothing is correct. The model's commitment to a sentential or propositional mode of knowledge representation renders it unable to account for explanatory understanding outside of that narrow context, and it generates a host of problems even within that context. Slow access, inarticulation of laws, and deductive inappreciation were discussed at the beginning of this essay. To these we may add the problems of explanatory asymmetry (Bromberger 1966, Teller 1974), irrelevant explanation (Salmon 1970), and accidental universals (Nagel 1962).

None of these difficulties attend the prototype activation model. Concerning the matter of access, relevant understanding is usually accessed in milliseconds. Concerning our inability to articulate laws, the PA model does not even suggest that we should be able to articulate them. For what gets accessed is not a stored universal conditional, but a complex prototype. Similarly, while our deductive incompetence is a problem for the D-N view, on the PA model deductive inference typically plays no role at all in the process of prototype activation. Moreover, as noted in section 4, etiological prototypes are in general temporally asymmetric. Explanatory asymmetries are thus only to be expected. Irrelevant explanations (such as appealing to 'hexed salt' in order to explain a sample's dissolving in water) appeal to strictly non-existent prototypes. And finally, what moves us to reject an impotent explanatory premise as a merely accidental universal is that fact that, despite its truth, it fails to express any of our learned prototypes.

What a well-turned deductive-nomological argument certainly can do is successfully evoke explanatory understanding in the hearer by provoking activation of the relevant prototype with some well-structured and highly salient verbal stimuli. D-N arguments are therefore entirely appropriate things to exchange in a great many explanatory contexts, especially in complex contexts. But they do not embody or account for the understanding itself.

It will be helpful at this point to summarize the presumptive advantages that the prototype activation model of explanation has over classical models, over the D-N model say, just to take a prototypical target (Table 1). On the left are listed familiar problems. On the right, I cite the PA model's peculiar virtue.

Some of the attempts made to patch the D-N model can be seen as well-motivated but opaque attempts to bridge the gap between the cognitive weakness of universally quantified conditionals and the cognitive muscle of genuine learned prototypes. In response to some of the standard difficulties, Brody (1972) suggested imposing a 'causal relation' condition and/or an 'essential feature' condition in addition to the usual D-N requirements. On the prototype activation model, the relevant difficulties are handled by something rather similar:

Table 1.

D-N problems	P-A virtues/solutions
1. Slow access to relevant explanatory beliefs	Access in milliseconds (no search at all)
2. Ignorance of Laws	We access prototypes, not laws
3. Deductive incompetence	Deduction is rarely, if ever, involved
4. Animals lack discursive laws	See 2. again
5. "Flagpole" problem	Etiological prototypes are temporally asymmetric
6. Contraposed laws problem	Etiological prototypes are temporally asymmetric
7. "Hexed salt" problem	No corresponding learned prototype
8. Accidental universals	No corresponding learned prototype
9. Logical diversity in types of explanation	Distinct major classes of explanatory prototypes
10. Why is simplicity/unity a virtue of explanations?	Unified partitions generalize better to novel cases

the existence within the creature of learned etiological prototypes and property-cluster prototypes. This may explain the appeal of Brody's proposals. However, these two kinds of prototypes constitute only two of five or six different kinds commonly deployed in explanation. It is a shortcoming of Brody's account that it encompasses, problematically, only these two kinds of explanation. Also, his account remains a restricted version of the D-N model. It is fundamentally sentential in character, and it does not escape the empirical objections voiced in section 1.

An important alternative to the D-N model is the *statistical relevance* or *SR* model (Salmon 1971). Perhaps the most salient and appealing feature of this model is its rejection of the requirement that the explanandum be inevitable or even probable in light of the explanans. Rather, understanding consists in grasping the statistics of the explanandum situation correctly, whatever those probabilities happen to be. Thus, it is said, one may properly explain my having the flu by pointing to my week of nursing a pair of flu-infected children, even though the probability of the former on the latter is less than ten percent in an adult.

Though many have contrary intuitions, I think this is a welcome modification. The explanans in such a case does throw light on the explanandum: one's understanding of the situation is significantly improved, despite the low probability at issue. In many domains, probabilities far below unity is all that the world ever provides. Explanatory understanding then appears on the thin side, but it does not disappear entirely.

For better or for worse, the prototype activation model seems committed to the same view. A prototype is by definition 'general', in the same sense in which a property is general – it has many instances. But that is the only sense in which it need be general. A prototype vector is not just a cluster of universally quantified conditionals in neural disguise. A prototype, you will recall, can represent a wide range of diverse examples. There need be no feature that is universal, or even nearly universal, to all examples in the class. The statistical distribution of relevant features, across the class comprehended by the prototype, may have almost any profile. But whatever that distribution is, it will be represented in the *salience* accorded each relevant feature in the similarity space configured at the relevant layer of hidden units. And it will likely be reflected in the subsequent cognitive and motor behavior that has been learned by the network. Activation of that vector, non-uniform though its portrait of the class may be, still represents the network's understanding of the input.

Though all of this may be true, I do not think it accounts for more than a small part of the explanatory appeal of the remark, "I nursed two flu-infected children for a week", relative to the question, "Why do you have the flu?" Rather, that explanans has the clout it does because it activates an etiological prototype concerning the typical transmission of viral infections, a prototype whose warrant in this case may be extremely high, despite the general low probability of adult illness given exposure to infected children. The warrant may be high because infection from those two children may be the only serious opportunity for infection you encountered, and we do demand some infective etiology for every case of flu. The importance of causal structures is also a major theme of Salmon's more recent work (1978, 1984) within the statistical perspective. From the PDP perspective, this represents a welcome recognition of the importance of etiological prototypes for explanatory understanding.

Overall, however, it also represents a vision of explanation that is confined to only the second of the five or six important kinds of prototypes outlined in section 4. Causal structures are relevant to some kinds of explanations, but certainly not to all. And despite its sharing some important commitments with the prototype activation model, the SR model remains firmly lodged within the sentential framework of folk psychology, and thus is heir to all of the empirical and psychological problems discussed in section 1.

A different and increasingly popular approach to explanation abandons hope of finding anything very interesting, well-defined, and unitary about what counts as an explanation. Van Fraassen (1977, 1980) has urged that whatever reduces someone's puzzlement can count, in that context, as an explanation, and it is evident that a considerable diversity of things can be expected to meet this vague and ultimately pragmatic condition. Here our attention is shifted away from understanding itself, and is focused instead on the polyglot variety of verbal or other acts that might in context produce it.

It is a consequence of the prototype activation model that the variety of things that can finally prompt a relevant prototype in a situation of puzzlement is endlessly various. There is nothing canonical about inputs that fail to activate a prototype, and there is nothing canonical about what subtle modifications or augmentations of those inputs will finally make them do so. A one-word hint, a salient gesture, an accidental analogy, almost anything might, in context, augment the available input so as finally to trip the activation of some appropriate prototype. "Why is everyone so tense around here this morning?", asks the office-manager of a random employee. A rolling of the eyes and a flicker of a glance toward the desk of the office troublemaker tells him all he needs to know.

Well and good. But endless variety in the immediate causal antecedents of explanatory understanding is wholly consistent with unity in the basic nature of understanding itself. The case for seeing such unity is detailed throughout this essay. The variety van Fraassen finds in explanatory contexts is entirely real, but I suggest it properly belongs only to the pragmatics of explanation.

Of course, van Fraassen's account is first and foremost an account of the pragmatics of explanation. But it directly suggests a correlative account of what explanatory understanding consists in. Taking his story at face value, explanatory understanding would presumably consist in the possession of an acceptable answer to a topic question, "Why P?", where the answer is acceptable because it entails or otherwise satisfactorily singles out the explanandum "P" from its local 'contrast class' of relevant alternatives.

I think this has to be wrong. Explanatory understanding is an ubiquitous phenomenon, an almost continuously unfolding feature of the bulk of any cognitive creature's ongoing conscious experience. It is not a specific answer that singles out a unique member from a specific contrast class of possible alternatives, contextually posed. Nor is it generally an answer to any question, explicit or implicit, at all. It is nothing remotely so linguistic, so stylized, or so occasional. In general, explicit questions arise only when explanatory understanding, for some reason or other, temporarily fails or eludes us. If we confine attention to such atypical cases, then van Fraassen's account of the pragmatics of 'cognitive recovery' may begin to look plausible. Yet even in these atypical cases it is plausible only for language using humans. Puzzlement and its resolution in non-human animals (that is, in most of the cognitive creatures on the planet) is not plausibly portrayed in van Fraassen's terms.

And not surprisingly. They are the same terms that got the D-N model in trouble. They portray knowledge as essentially propositional in character.

Accordingly, what van Fraassen has given us is, at most, an account of the structure and rules of a certain game, an essentially linguistic game, a game that is occasionally played by the older members of a single species of animal on those exceptional occasions when explanatory understanding for some reason eludes them. It does not provide, nor does it imply, a satisfactory account of explanatory understanding itself.

Perhaps because of his general commitment to Constructive Empiricism – which sees bare descriptive empirical adequacy as the only genuine epistemic virtue – van Fraassen despaired too quickly of finding anything unitary, interesting, and *epistemologically significant* in the general topic of explanation. On the prototype activation model, however, explanatory understanding emerges as being on a cognitive continuum with perceptual recognition, a genuinely epistemological matter if ever there was. Additionally, as we saw at the end of the preceding section, explanatory unification is an important and understandable determinant of how successfully one is able to generalize past experience to new cases, which is also an epistemological rather than a merely pragmatic matter. Let us look more deeply into this topic.

I have made conceptual unification an important element of the account proposed. But the present account is not the only one that accords a central importance to explanatory unification. Friedman (1974) and Kitcher (1981, 1989) have both urged that the virtue of an explanation is a function of whether it is, or promises to be, an element or instance of a unified account of a broad range of phenomena. Explanatory virtue, according to them, is not just a matter of the relation between the local explanans and the local explanandum. It is also a matter of the global virtues of the framework of which the local explanans is but one application. While each offers a different account of what the explanatory unification of our knowledge consists in, they are agreed in making participation in such unity the dominant virtue of any explanation.

On the view of explanation here being defended, they are almost certainly right. Some will resist the elevation of unity to this level of importance, arguing that explanatory unity is a purely aesthetic or pragmatic virtue, valued by humans perhaps, but strictly irrelevant to truth (van Fraassen 1980). Yet we have already seen in section 5 how a 'superempirical' virtue like conceptual unity has inevitable consequences for the 'empirical adequacy' of a cognitive configuration, as displayed in its ability to discriminate correctly novel instances of a learned category. The lesson here repeats an earlier lesson (urged in Churchland 1985a): we should be very reluctant to assent, as van Fraassen assents, to any distinction drawn between the 'genuinely empirical' and hence 'truth-relevant' virtues of a theory, and its 'merely superempirical' and hence 'truth-irrelevant' virtues. The distinction is without integrity.

We can push the general lesson of section 5 even farther. If perceptual recognition and explanatory understanding are really instances of the same form of cognitive achievement, as I have suggested repeatedly in this essay, then it is proper to regard perceptual recognition itself as being just a case of explanatory understanding at the sensory periphery. On this view, the factors that make for worthy modes of explanatory understanding are the very same factors that make for worthy modes of perceptual apprehension of the world in the first place.

This vital point is the motivating theme of my 1979 book. The insight derives ultimately from C. S. Peirce, who explicitly voices the position on perceptual recognition here being defended: ". . . perceptual judgment is the limiting case of abductive judgment." (in Hartshorne and Weiss 1935: vol. 5, para. 186). We can now sustain Peirce's insight (and Gregory's and Rock's) with a revealing and neurally grounded account of what abductive judgment really consists in. It consists in prototype activation. With choices at the empirical periphery thus hopelessly hostage to explanatory values, we can now see that the virtue of explanatory unity must play a dominant role in any adequate epistemology. It cannot be demoted to a merely pragmatic status, as it is in the austere ideology of Constructive Empiricism.

The PDP perspective also allows us to say something useful about the conflict between the Gibsonian 'ecological' approach to perception and the 'New Look' approach to the same topic. The first point to make is that the Gibsonians appear to be right on the following matter. Perceptual processing is impressively fast and direct: inference appears to play no role at all. On the other hand, the New Look theorists appear to be right on the following point. Our perceptual processing is profoundly theory laden, in that it is dictated largely by the accumulated knowledge – the configuration of weights – acquired in the course of past experience. There is no longer any conflict between these two convictions. They are both a part of the PDP perspective.

(A notable view that does suffer, however, is Fodor's view (1984), which attempts to embrace the worst of both positions. Fodor and Pylyshyn (1981) have argued against the Gibsonians that the perceptual process is extensively inferential, and Fodor (1984) has argued against me (1979), and other New Look sympathizers, that nevertheless the process is basically theory-neutral.)

Let us return to Kitcher's account of explanation. It does more than compellingly portray unity as the pre-eminent cognitive and explanatory virtue that it is. It also discovers and profitably deploys the idea of prototypes. Kitcher takes an important step away from the traditional D-N picture by introducing the notion of an abstract *argument pattern*. These are highly abstract logical structures, characteristic of the particular scientific theory being exploited, structures that will yield a great variety of relevantly specific arguments when instantiated to suit a specific explanatory problem. One is reminded here of Kuhn (1962). But unlike Kuhn, Kitcher explains the matter in great detail, and shows how an impressive amount of light can be thrown on many chronic problems if we adopt his point of view.

From the PDP perspective, Kitcher's achievement must be judged genuine. The deployment of prototypes is what is central to explanatory understanding, and for maximal virtue it is vital that they be part of a unified cognitive configuration. The problem with Kitcher's story is that it is still much too narrowly *linguistic* in its conception of knowledge representation, and it encompasses only a small subset of the full range of different types of explanation. These are serious defects in the story. Even so, I would much rather highlight its equally prominent virtues, since I would like to claim that my own account successfully reduces it. The incorporation of the general features of Kitcher's story further illustrates the conceptual unity that the prototype activation model brings to the topic of explanation.

If the conceptual unity it brings is genuine, then it ought to permit the recognition and understanding of hitherto unrecognized types of explanation. There is some sign of this. The belated recognition of legal, moral, and social etiquette explanations was a novel experience for me, I am somewhat embarrassed to say, and one might suggest aesthetic explanation as a further category. Perhaps others will see more.

Finally, let me close this essay by adverting to a highly general virtue of the prototype activation model. I am compelled to cite it since, as I look back over this essay, I am distressed at how fragile is the account proposed, and how sketchy are the few details provided. What makes me hopeful in the face of this is the fact that the prototype activation model of explanation meets its own primary condition of virtue, a condition of independent plausibility. Specifically, that model brings a welcome and revealing unity into a stubborn diversity of explanation types, and the model is itself an integral part of a highly unified background conception of cognitive activity, one that encompasses with some success the general structure and activity of biological brains, and the structure and cognitive behavior of a new class of artificial computing systems. For this reason, if no other, we should be moved to explore it further.

Department of Philosophy, B-002 UCSD,
La Jolla, CA 92093

NOTE

* This paper was first presented at the Jagellonian University, Krakow, and first published as chapter 10 of P. M. Churchland (1989).

REFERENCES

Brody, B. A. (1972), 'Towards an Aristotelian Theory of Scientific Explanation', *Philosophy of Science 39*.

Bromberger, S. (1966), 'Why-Questions', in R. Colodny, ed., *Mind and Cosmos* (Pittsburgh: Pittsburgh University Press).

Churchland, Paul M. (1979), *Scientific Realism and the Plasticity of Mind* (Cambridge: Cambridge University Press).

Churchland, Paul M. (1981), 'Eliminative Materialism and the Propositional Attitudes', *Journal of Philosophy LXXVIII*, no. 2.

Churchland, Paul M. (1985), 'The Ontological Status of Observables: In Praise of the Superempirical Virtues', in Churchland, P. M., and Hooker, C. A., eds., *Images of Science* (Chicago: University of Chicago Press).

Churchland, Paul M. (1986), 'Some Reductive Strategies in Cognitive Neurobiology', *Mind 95*, no. 379.

Churchland, Paul M. (1988a), 'Folk Psychology and the Explanation of Human Behavior', *Proceedings of the Aristotelian Society: Supplementary Volume LXII*.

Churchland, Paul M. (1988b), 'Simplicity: The View from the Neuronal Level', in Rescher, N., ed., *Aesthetic Values in Science: Pittsburgh Colloquium in the Philosophy of Science* (forthcoming).

Churchland, Paul M. (1989), 'On the Nature of Theories: A Neurocomputational Perspective', in W. Savage, ed., *Minnesota Studies in the Philosophy of Science: Vol. 14* (Minneapolis: University of Minnesota Press).

Fodor, J. A. (1984), 'Observation Reconsidered', *Philosophy of Science 51*, no. 1.

Fodor, J. A., and Pylyshyn, Z. (1981), 'How Direct is Visual Perception? An Examination of Gibson's 'Ecological Approach'', *Cognition 9*, no. 2.

Friedman, M. (1974), 'Explanation and Scientific Understanding', *Journal of Philosophy LXXI*.

Gorman, R. P., and Sejnowski, T. J. (1988), 'Learned Classification of Sonar Targets Using a Massively-Parallel Network', *IEEE Transactions: Acoustics, Speech, and Signal Processing 36*, 1135–40.

Gregory, R. (1966), *Eye and Brain* (London: McGraw Hill).

Gregory, R. (1970), *The Intelligent Eye* (New York: McGraw Hill).

Harman, G. (1965), 'Inference to the Best Explanation', *Philosophical Review 74*.

Harman, G. (1973), *Thought* (Princeton: Princeton University Press).

Hartshorne, C. and Weiss, P. (1935), *Collected Papers of Charles Sanders Peirce* (Cambridge: Harvard University Press).

Hempel, K. (1965), 'Studies in the Logic of Explanation', in Hempel, K., *Aspects of Scientific Explanation* (New York: The Free Press).

Hesse, M. (1966), *Models and Analogies in Science* (Notre Dame: Notre Dame University Press).

Kitcher, P. (1981), 'Explanatory Unification', *Philosophy of Science 48*, no. 4.

Kitcher, P. (1989), 'Explanatory Unification and the Causal Structure of the World', in Kitcher, P., ed., *Scientific Explanation: Minnesota Studies in the Philosophy of Science, Vol. 13*.

Kuhn, T. S. (1962), *The Structure of Scientific Revolutions* (Chicago: University of Chicago Press).
Lakoff, G. (1987) *Women, Fire, and Dangerous Things* (Chicago: University of Chicago Press).
Lehky, S. and Sejnowski, T. J. (1988a), 'Computing Shape from Shading with a Neural Network Model', in Schwartz, E., ed., *Computational Neuroscience* (Cambridge: MIT Press).
Lehky, s. and Sejnowski, T. J. (1988b), 'Network model of shape-from-shading: neural function arises from both receptive and projective fields', *Nature 333* (June 2).
Minsky, M. (1981) 'A Framework for Representing Knowledge', in Haugeland, J., ed., *Mind Design* (Cambridge: MIT Press).
Nagel, E. (1961), *The Structure of Science* (New York: Harcourt, Brace & World).
Putnam, H. (1975) 'Is Semantics Possible?', and 'The Meaning of 'Meaning'', in Putnam, H., *Mind, Language and Reality* (Cambridge: Cambridge University Press), 139.
Rock, I. (1983), *The Logic of Perception* (Cambridge: MIT Press).
Rosch, E. (1981) 'Prototype Classification and Logical Classification: The Two Systems', in Scholnick, E., ed., *New Trends in Cognitive Representation: Challenges to Piaget's Theory* (New Jersey: Lawrence Erlbaum).
Rosenberg, C. R. and Sejnowski, T. J. (1987), 'Parallel Networks That Learn To Pronounce English Text', *Complex Systems 1*.
Rumelhart, D. E., Hinton, G. E., and Williams, R. J. (1986a), 'Learning representations by back-propagating errors', *Nature 323*.
Rumelhart, D. E., Hinton, G. E., and Williams, R. J. (1986b), 'Learning Internal Representations by Error Propagation', in Rumelhart, D. E., and McClelland, J. L., eds., *Parallel Distributed Processing: Explorations in the Microstructure of Cognition* (Cambridge: MIT Press, 1986).
Salmon, W. (1970), 'Statistical Explanation', R. Colodny, ed., *The Nature and Function of Scientific Theories* (Pittsburgh: Pittsburgh University Press).
Salmon, W. (1971), *Statistical Explanation and Statistical Relevance* (Pittsburgh: Pittsburgh University Press).
Salmon, W. (1978), 'Why Ask, 'Why'?', *Proceedings and Addresses of the American Philosophical Association 51*.
Salmon, W. (1984), *Scientific Explanation and the Causal Structure of the World* (Princeton: Princeton University Press).
Schank R. and Abelson, R. (1977), *Scripts, Plans, Goals, and Understanding* (New Jersey: John Wiley and Sons).
Sturgeon, N. (1985), 'Moral Explanation', in Copp, D. and Zimmerman, D., eds., *Morality, Reason and Truth*, supplementary volume 24, *The Southern Journal of Philosophy*.
Teller, P. (1974), 'On Why-Questions', *Nous VIII*.
Van Fraassen, B. (1977), 'The Pragmatics of Explanation', *American Philosophical Quarterly 14*.
Van Fraassen, B. (1980), *The Scientific Image* (Oxford: Oxford University Press).
Wittgenstein, L. (1953), *Philosophical Investigations* (Oxford: Basil Blackwell), sec. 66–77.

TERESA GRABIŃSKA

THE AMBIGUITY OF THE LAKATOS-ZAHAR RECONSTRUCTION OF CLASSICAL AND RELATIVISTIC PROGRAMMES

1. INTRODUCTION

Since the triumphant procession of the relativistic programme many works criticizing this approach appeared in physics. Most frequently they did not manage to cope with the requirements and provisions imposed by the physicists and methodologists. We shall consider the Ives' approach to the Einsteinian revolution, which is a purely operationist kinematics.

The kinematics of H. E. Ives[1] can be characterized as competitive with the special relativity theory (STR) of Einstein, although it is posterior to the latter (it came into being in 1937–53). These two theories refer to the same problem, the question of relative motion of reference systems moving with uniform rectilineal motion, and formulae of the theory of Ives (TI) correspond to the STR formulae. The theories differ from each other with respect to absolute elements: *TI* is a theory which assumes the existence of an absolute reference system – ether, while STR introduces the complete relativity of reference systems and at the same time the absolute velocity standard of velocity-*c*.

In this paper, we shall compare the assumptions of the two kinematical theories, STR and *TI*, in the scheme of the Lakatos-Zahar methodology of research programmes.[2] This comparison provides many results[3] e.g.: (1) it helps to formulate an opinion about Einstein's need to introduce his controversial light postulate, (2) it allows us to trace mutual relation of operational assumptions in the theory with regard to the other assumptions of a heuristic nature, (3) it helps us to realize the necessity to introduce abstract constructs of time in STR. The comparison of these theories also throws light on the correspondence relation of the theories: namely *TI* gives formulae for STR in the limit of a non-realistic procedure of measurement.

Comparative analysis of the two theories gives results for the methodology of scientific research programmes: it provides examples of an ambiguous classification of theory relative to a given programme and also a non-absolute character of the programme division into hard core and heuristics.

J. Misiek (ed.), The Problem of Rationality in Science and its Philosophy, 109–117.

2. EINSTEIN'S RELATIVITY PROGRAMME

The change of the modern physics paradigm from classical Newtonian physics had been discussed since 1905 when Einstein published his famous work[4] introducing a new kinematics, the so-called relativistic kinematics as compared to the Newtonian kinematics. The new kinematics initiated a rapid development of relativization in physics and new research programme. The old programme, so-called Lorentz programme, stopped making new contributions to the development of physics but this did not mean that classical theories have no longer been the basis of many modern technical and applied sciences. However, although the absolutist understanding of time and space became an anachronism, measurement procedures remained the "classical" ones which presented many troubles not only in quantum physics but also in the relativistic theories.

Zahar in his well known work[5] attempted to give an answer to the question, in what sense was the Einstein programme dominant in the relation to the Lorentz programme? The question referred to the period directly following the issue of Einstein's pioneer work. Zahar answered that the heuristics of the relativity programme attracted the attention of physicists at the beginning of the twentieth century: The new heuristics of relativization of physics came to be decisive for the abandonment of Lorentz's programme.

For further consideration, let us reconstruct briefly the features of the two research programmes. In the Lorentz programme, the heuristics H_c consists of the ether hypothesis, that is the assumption of the distinctive role of an absolute reference system: namely, all physical phenomena are the result of interactions transferred through the actual actions of this absolute reference system, the luminiferous ether[6].

The hard core HC_c of the classical programme consists of the first two Newtonian laws of mechanical motion, Maxwell's equations for the electromagnetic field, and the Galilean relativity principle. The Lorentz programme is the conjunction of HC_c and the equation for the so-called Lorentz force; expressions obtained by Lorentz can be derived on the basis of these assumptions. The reconciliation of the ether heuristics and Galileo's principle of relativity with Maxwell's equations for the problem of particle motion in an electromagnetic field led Lorentz[7] to find the transformations which later were named by Poincaré[8] for Lorentz. Then the conjunction of HC_c and the Lorentz transformations generated the theory of corresponding states TCS[9]; the conjunction of HC_c, TCS, and the hypothesis of a molecular forces[10] (which is not an *ad hoc* hypothesis) implied the rigid rod contraction and the clock dilatation. Developed in such a way Lorentz programme was equivalent to STR from both the theoretical and the empirical points of view.

Einstein's heuristics H_R in Zahar's reconstruction[11] is as follows: (1) science should present us with a coherent, unified, simple, organically compact picture of the world; (2) the old theory, which does not explain symmetrical observational coincidences as the manifestations of deeper symmetries, or which does not anticipate facts known from observation, should be replaced by a new theory which has these properties. Modern theoretical physics makes abundantly use of this heuristics in the programme of the uniform description of physical interactions: the mathematical elegance of this description is included e.g. in the symmetry group of equations. The adoption by physicists of the Einsteinian heuristics allowed them to develop new physical theories, and this heuristics still seems to be promising for the unification of physical interactions and fields, in the search for uniform field theoretical descriptions for elementary particles and critical phenomena in the physics of phase transitions, in the development of physical cosmology to explain the time-space evolution of the world and the origin of elementary particles (and chemical elements), and of grain structure of the Universe including galaxies in the remotest time.

The unification of Newtonian mechanics and the Maxwell-Lorentz electrodynamics within the new relativistic kinematics was the main achievement of Einstein's first work of 1905 dealing with STR. The fundamental significance of STR did not consist in the presentation of new formulae and it should be mentioned that before 1905 Lorentz and Poincaré derived all the transformation expressions appropriate to STR, and indeed the Lorentz TCS was, as Poincaré showed, theoretically and empirically equivalent to STR. In spite of Einstein's detailed studies of Lorentz's works of 1892 and 1895, the STR postulates and the heuristics of the new programme were different from those characteristic of the programme of classical physics.

Heuristic hints H_R have been presented while the hard core of STR consists of two postulates: *P(1)* – special relativity principle – "if *K* is an inertial system then every other system *K′* which moves uniformly and without rotation relatively to *K*, is also an inertial system; the laws of nature are in agreement for all inertial systems", *P(2)* – light postulate – in a vacuum, light is propagated with velocity *c* in every inertial system[12]. The postulate of space isotropy, the so-called principle of direction relativity, is in addition to these two postulates.

While *P(1)* does not stir up controversies (it is a generalization of the Galilean relativity principle modified later by Poincaré in 1904[13]), on the other hand the second postulate *P(2)* has been the subject of investigation until now regarding first its genesis in original Einstein's work of 1905, and second its experimental justification. Einstein himself claimed that *P(2)* was not of an inductive nature, i.e. the result of the Michelson-Morley experiment did not have an essential influence on his expression of the fundamental postulate *P(2)*, but he also presented different opinions as to its nature and origin.[14]

3. RECONSTRUCTION OF IVES'S RESEARCH PROGRAMME

If now we consider[15] the kinematic theory of Ives, it is possible to distinguish the following postulates in the hard core (HC_I): (1) postulate of the independency of light velocity from the velocity of the source, and its numerical value *c* in the absolute reference system only, (2) postulate of contraction and dilatation of measurement instruments in motion, (3) principle of relativity of Poincaré stating that the laws of physical phenomena should be the same whether the observer is at rest or he moves rectilineally and uniformly. However, it should be added that such reconstruction of the hard core of Ives's programme, as mentioned above, refers to the development of this programme up to 1949. Later, in the new hard core HC'_I, Ives "dethroned" postulate (2) about the change of instruments in motion deriving it from the remaining two postulates (1) and (3).

The Ives's heuristics H_I can be reconstructed in the following way: (1) the luminiferous ether constitutes the absolute reference system for all motions (interactions) and for physical measurements, (2) physical equations should be built so as to reflect the behaviour of measurement instruments and all the quantities in physical equations should have an operational meaning.

In the age of the modern development of physics, an attitude admitting the logical basis for the existence of an absolute reference system is regarded as inconsistent with the canons of modern physics. Ives realized this very well when in the course of many years he developed his theory of the motion, which was competitive to STR and based on the ether heuristics. Particularly he referred to this problem in his last work[16] published before his death. The radiation reaching an observer from distant stars was the realization of ether in the Ives theory. The Ives kinematics required a cosmic background (thus it was not completely local), while in STR such a global physical system was redundant, useless. Ether played the function of transmitting electromagnetic signals with a velocity independent of its material source. Instruments which were stationary relative to the ether provided the Newtonian understanding of velocity.In the case of real instruments it was assumed that they are in motion relative to the ether (there was no stationary instrument relative to the ether).

The second heuristic component of the Ives programme recommended construction of the physics equations so that the quantities occurring in them should have an operational interpretation i.e. the quantities should be defined on the basis of real measurement procedure. According to Ives, the correctly defined measurement procedures are an integral part of the theory in addition to the formal equations.

4. DIFFICULTIES WITH PUTTING THE IVES'S THEORY INTO THE LORENTZ OR EINSTEIN RESEARCH PROGRAMMES

The theory of Ives[17] contains heuristic hints which direct us to refer all motions (and measurements) to the absolute frame of reference. This feature suggests ties with the Lorentzian research programme. However the task of classification is complex.The original form of *TI* (with HC_I) presents difficulties for its reconstruction in the programme of classical physics because the rod contraction and clock dilatation have in it a status of hard core postulate whereas in the classical programme they are derived from axioms of HC_c. The second version of *TI* (with HC_I) also cannot be compared with the Lorentzian programme because derivation of rod contraction and clock dilatation in *TI* is not equivalent to the origin of this property in the programme of classical physics. As pointed out by Zahar, it is the consequence of a hypothesis about molecular forces and is of a dynamic-kinematical nature.

The transformation formulae between inertial reference systems are the same in *TI* and STR but *TI* cannot be considered as a component of the relativistic programme: *TI* differs essentially from STR in heuristics, which in the case of *TI* is an absolutist one. Finally *TI* does not bring any innovative features in the axioms, heuristics and consequences, which would allow us to consider it to be a separate research programme. So it is possible to ask a question about the universality of the reconstruction of knowledge growth in the spirit of the Lakatos' methodology. The fragment of physics history presented here does not let us take a strong line regarding this important problem. The general answer consists in this: undoubtedly it is possible to reconstruct research programmes as sequencies of theories built according to the postulates which comprised the hard core, by means of heuristic hint, however it seems that the range of these reconstructions is limited.

The more detailed reasons for the limitation of the effectiveness of Lakatos's reconstruction in the case of *TI* versus the Lorentz and Einstein programme are the following:

1. Our case of three competitive theories considered here has difficulties when we try to find clear criteria of classification into separate research programmes. It is one more link between classical and relativistic programmes. *TI* does not fulfil all the requirements of H_R but more consistently satisfies the operational requirement of STR which is illegitimately neglected in the Zahar's reconstruction. Taking explicit account of the procedures of clock synchronization shows that *TI* is in some sense more "relativistic" (consistent with the relativity principle for motions and measurements) than STR is.[18]

2. The chain of theories derived from few axioms presents a very crude understanding of knowledge growth. Every theory has its own language which has many relationships to the language of previous theories. The comparison of *TI* and STR shows that the concepts in STR are in fact more classical (e.g. velocity) than the respective ones of TI.

3. The transformation formulae of *TI* have the same simple form as the formulae of STR and the Lorentz TCS. Similar to the Lorentz theory, the kinematical quantities in them have a complicated interpretation in inertial frames of reference. From this point of view *TI* may be classified as a branch of the Lorentz programme. However in this case the conception of research programme should be generalized and branches separately developing within the research programme should be admitted. So a hierarchy of research subprogrammes (in the form of a tree) would better reflect the real development of science than the Lakatosian sequence of theories in a single research programme.

4. Of course, the comparison of the Einstein and Lorentz theories (e.g. in the form presented by Zahar) throws a new light on the structure of each of them. However other comparisons (as in the case of *TI*) are also very fruitful although they do not lead to a clear classification into research programmes. The comparison presented shows that all three theories contain absolute theoretical concepts. In the Lorentz theory and *TI*, they are expressed explicitly: "there is an absolute reference system" whereas in STR they are hidden in the light postulate. The constancy of light velocity plays the same role as ether does[19]: either

(a) it is assumed that absolute reference system exists in regard to which all the velocities are determined (as well as measurements are performed) and in which exceptionally $c_o = c_b = c$, where c_o is light velocity "to", and c_b is light velocity "from" (the case of *TI* and Lorentz theory), or (b) only kinematically equivalent systems are accepted, and $c_o = c_b = c$ in each reference system is assumed to be a theory postulate (the case of STR).

The mentioned inadequacy of the Lakatos-Zahar reconstruction allows us to prepare another logical reconstruction of the origin of the relativistic research programme with respect to the classical programme. Its starting point is a coordination between the theoretical quantities and the quantities measured by experiment. The distinguishing role of this operational criterion in the situation when the theoretical apparatuses and empirical bases are identical, may be decisive for the form of the theory. In the light of new reconstruction, the operational equivalence of the postulates of ether existence and the Einstein light postulate arises sharply.

5. If we are to judge the arguments about the position of the proposition of ether existence and the role of ether in the hard core or the heuristics of research programmes of Lakatos-Zahar, we should rather concur in Miller's approach[20] (a proposition about the ether contained in hard core) than the opposed attitude of Zahar. The thesis that some basic propositions can serve the role of axioms and at the same time that of heuristic principles for the research programme can be a compromise issue in the light of arguments about the function of the proposition about the ether in research programme. The proposition about the ether can have the form of a postulate from the hard core in the case of a programme with an absolute reference system: (1) the ether exists as absolute reference system for all the motions (and measurements), and in ether the light velocity is c; or it has the form of metaprinciple from the programme heuristics[21]: (2) all the physical phenomena and interactions must be referred to the ether. Proposition (1) about the ether plays a role similar to the two basic fundamental postulates *P(1)* and *P(2)* in STR. Proposition (2) is the clue which recommends to consideration of the true essence of physical phenomena in the absolute reference system, the ether. The necessity of compliance with proposition (1) requires the adoption of different heuristic hints in the process of programme development than those included in proposition (2).

6. The relativistic heuristics H_R, as it is reconstructed by Zahar does not present any characteristic attributes of the very relativistic programme itself: the requirements of mathematical simplicity and an unified description of physical phenomena are heuristic components of all of modern physics and also can be seen in the Lorentz theory as well in TI. The more original Einstein metaphysics can be displayed[22] but still it does not allow us to appraise research programmes in Zahar's fashion.

Technical University of Wrocław
Poland

NOTES

[1] Ives H. E., "Light Signals on Moving Bodies as Measured by Transported Rods and Clocks", *JOSA 27* (1937), 263–273; "Apparent Lengths and Times in Systems Experiencing the FitzGerald-Larmor-Lorentz Contractions", *JOSA 27* (1937), 310–313; "Derivation and Significance of the So-called Chronotopic Interval", *JOSA 29* (1939), 294–301; "Derivations of the Lorentz Transformations", *Phil. Mag. 36* (1945), 392–403; "The Measurement of the Velocity of Light by Signals Sent in the One Direction", *JOSA 38* (1948), 879–884; "Lorentz Type Transformations as Derived from the Performable Rod and Clock Operations", *JOSA 39* (1949), 757–761; "The FitzGerald Contraction", *Scientific Proc. Royal Dublin Soc. 26* (1952), 9–26; "Genesis of the Querry 'Is there an Ether?'", *JOSA 43* (1953), 217–218.

[2] Lakatos I., "History of Science and its Rational Reconstruction", *Boston Studies in the Philosophy of Science 8* (1971), 91; Lakatos I. and Zahar E., "Why did Copernicus' Programme Supersede Ptolemy's?", in *The Copernican Achievement*, ed. R. Westerman, The University of California Press, Los Angeles, 1976, 354–383; Zahar E., "Why did Einstein's Programme supersede Lorentz's?", in *Method and Appraisal in the Physical Sciences*, ed. C. Howson, Cambridge University Press, 1976, 211–275.

[3] Grabińska, T., "The Analysis of Operationistic Assumptions of the Ives' Kinematics and Relativistic Kinematics", (in Polish), *Z Zagadnien Filozofii Przyrodoznawstwa i Filozofii Przyrody VIII* (1986), 7–34; "Relativity and Space–Time Measurements", *Astroph. Space Sci. 191* (1992) 23–42.

[4] Einstein, A., "Zur Elektrodynamik bewegter Körper", *Annalen der Physik 17* (1905), 891–921.

[5] Zahar E., 1976, *op. cit.* in ref. 2.

[6] Compare the discussion about the possibility of placing the assumption about ether either in heuristics or in hard core of the research programme: Miller A.I., "On Lorentz's Methodology", *Brit. J. Phil. Sci. 25* (1974), 29–45; Zahar E., "Einstein's Debt to Lorentz", *Brit. J. Phil. Sci. 28* (1978), 49–60.

[7] Lorentz H. A., "La Théorie Electromagnétique de Maxwell et Son Application aux Corps Mouvants", *Arch. Nederland, Sci. Exact. Nat. 25* (1892), 363–642; "The Relative Motion of the Earth and the Ether", *Vers. Kon. Akad. Wetensch. Amsterdam 1* (1892), 74–78.

[8] Poincaré H., "Sur la dynamique de l'electron", *Comptes Rendus 140* (1905), 1504–1508; *Rendiconti Circolo Matem. Palermo 21* (1906), 129.

[9] Lorentz H. A., "Théorie simplié des Phènomènes Electriques et Optiques dans des Corps en Mouvement", *Versl. Kon. Akad. Wetensch, Amsterdam 7* (1899), 507–523.

[10] Zahar E., 1976, *op. cit.* in ref. 2, pp. 229–230.

[11] *Ibid.*, pp. 237–239; c.f. Zabierowski M., "Two approaches to cosmology", (in Polish), *Zagadnienia Naukoznawstwa XXIII* (1987), 482–489, where a different heuristics of the relativistic programme, is presented.

[12] Einstein A., *op. cit.* in ref. 4; *The Meaning of Relativity*, Princeton University Press, 1955.

[13] Poincaré H., "L'état actuel et l'avenir de la physique mathematique", La *Revue de Idees* (1904), 804.

[14] Shankland R. S., "Controversions with Albert Einstein", *American J. Phys. 31* (1963), 47–57, but Einstein also presented different opinions on the origin of *P(2)*, e.g. Einstein A., "How I Created the Theory of Relativity", *Physics Today* (1982), August, 45–47.

[15] Grabińska, T., 1986, 1992, *op. cit.* in ref. 3.

[16] Ives, 1953, *op. cit.* in ref. 1.

[17] Ives, 1937, 1939, 1945, 1948, 1949, *op. cit.* in ref. 1.

[18] Grabińska, 1986, 1992, *op. cit.* in ref. 3.
[19] Grabińska, 1986, 1992, *op. cit.* in ref. 3.
[20] Miller, 1974, *op. cit.* in ref. 6.
[21] Zahar, 1976, *op. cit.* in ref. 2.
[22] Zabierowski, 1987, *op. cit.* in ref. 11.

WŁADYSŁAW KRAJEWSKI

APPROXIMATIVE TRUTH AND DEPTH AS THE MAIN AIMS OF SCIENCE

A discussion with John Watkins

1. IS TRUTH THE AIM OF SCIENCE?

I fully agree with John Watkins that the aim of science must be feasible and must "involve the idea of truth", that science "aspires after truth". I also agree that the Bacon-Descartes ideal of science – the requirement of absolute truth and certainty – is no longer valid (Watkins, 1984, Ch. IV). What then to do with the idea of truth?

Watkins gives the following solution. A scientific statement (hypothesis) adopted by a scientist should be "possibly true" for him in a sense that, "despite his best endeavours, he has not found any inconsistencies in it or between it and the evidence available to him" (pp. 155–56). However this aim is not feasible!

Many philosophers of science show that inconsistencies between a theory and available evidence are always – or almost always – present, that a new theory "floats in a see of anomalies", it is "born refuted" and "dies refuted". Watkins strives to show that not all theories are "born refuted" (pp. 330–334). Maybe, but it is enough for us that some of them are "born refuted". Other anomalies appear later – the theory is, however, not abandoned for this reason. Even when a theory is replaced by another one, we know that the old theory usually is not abandoned but preserved in a narrower domain, where it yields a good approximation to reality. Nobody refutes a theory when he finds an inconsistency . . . . Watkins postulate is not realistic, not feasible. Neither truth *tout court*, nor "possible truth" can be considered as the aim of science. We must search for another solutions.

2. APPROXIMATIVE TRUTH

I think the main fault of the Watkins' book, like of the majority of books in the philosophy of science, is neglect of the concept of approximation. True,

J. Misiek (ed.), The Problem of Rationality in Science and its Philosophy, 119–123.

he considers the concept of exactitude but not in a context of approximation and truth. In my opinion, the problem of the degree of the approximation to reality is the most important problem connected with the truth in science. Physicists have not doubts that all or almost all laws and theories give not the accurate but only approximate description of reality. But philosophers usually do not notice this. They most often give examples of qualitative statements which are indeed simply true or false. They also speak about probabilities of such statements. However, it is not qualitative statements that are important for physics (and not only for physics). Quantitative laws and theories form its core. They are usually not only probable but also approximate.

As known, Karl Popper has introduced the concept of verisimilitude (VS). According to him, in the course of the growth of science the VS of its theories increases, they are closer to truth, they approximate reality better and better. Popper gave a definition of VS but, as other authors showed, his definition does not withstand criticism. Nevertheless, in the main Popper's idea is valid and very important. The increasing VS or approximation is the aim of science, the criterion of its progress. Watkins mentions such opinions but refuses to include VS into his list of the aims of science: since there is no good criterion of VS it cannot "serve as a guide in the making of choices between competing hypotheses" (p. 280). However, some other aims adopted by Watkins also do not have commonly admitted criteria. The choice between competing hypotheses is always a difficult task and its solution is to great extent based on the intuition of scientists.

I prefer to speak about approximative truth instead of VS. I consider it not as a kind of falsity but as a kind of truth, and more, the kind of truth most important for contemporary science.

Popper tried to define VS by means of sets of true and false consequences of a theory. I think it was a wrong way. The only right way is the evaluation of the approximation of theories to reality, the introduction of concepts of approximative truth and its truth-content. I gave definitions of these concepts in my (1977). I shall not present them here all, the more because they need some improvement.

According to my conception, the truth-content (TrC) of scientific theories increases in the course of the growth of science. E.g., The Copernican Helio-Centric system (HC) was approximately true but its TrC was not high, its approximation was very rough. Kepler Laws (KL) gave a better approximation, their TrC was higher. Newtonian Classical Mechanics (CM), applied to the description of the planetary system, gave further progress, taking into account the perturbations of planetary orbits. Einstein's Special Relativity (SR) and

General Relativity (GR) mark the next steps in the increasing of truth-content, in a more exact description of of the motion of planets. We may write

$$\mathrm{TrC(HC)} < \mathrm{TrC(KL)} < \mathrm{TrC(CM)} < \mathrm{TrC(SR)} < \mathrm{TrC(GR)} < \ldots \quad (1)$$

One of the main aims of science is the best approximation to reality, the approximative truth with the best TrC, i.e. a theory which is possibly close to the exact description of certain domain of reality. However, the latter is probably unattainable. The process of approximation to reality seems to be infinite.

3. DEPTH

The second main aim of science is maximum depth. Here I agree with Watkins who speaks about the "depth-pole" of the ideal of science. Science penetrates into the internal structure of matter, the internal mechanisms of phenomena. This penetration is gradual. We may say that the essence of phenomena has many layers, maybe an infinite number of them. The deeper we penetrate into the essence, the better is the explanation of phenomena that we can obtain. Simple examples will illustrate this.

The wave theory of sound disclosed its essence; the molecular-kinetic theory, the essence of heat; the electron (and ion) theory, the essence of electrical current; the conception of metabolism, the essence of life; the conception of genetic information, the essence of heredity, etc. We see that in this approach the term 'essence' ceases to be metaphysical and becomes scientific. It is a synonym of internal structure of an object or internal mechanism of an phenomenon. We have to do not with the ultimate essence allegedly disclosed by philosophy but with an essence disclosed by science. We may call this approach, after Popper, 'modified essentialism' (Popper 1972, p. 195).

We mentioned various layers of essence and their gradual penetration. The following sequence of theories can serve as an example.

Empirical laws of chemical affinity, proportions of elements in compounds (ChA) describe the 'surface' regularities of chemical phenomena, and they give a very shallow explanation of these phenomena. The following theories penetrate subsequent layers of the essence of chemical phenomena and yield deeper and deeper explanation of them: the atomic-molecular theory of matter (AM) with the concept of valency of atoms, the theory of electron shells in the atom (EA) with the role of external shells, the theory of proton-neutron structure of the atom nucleus (PN) with the concept of the electric charge of the nucleus.

When we designate the depth of a theory T by D(T) we have the following sequence:

$$D(ChA) < D(AM) < D(EA) < D(PN) < \ldots \qquad (2)$$

I do not propose any measure of depth (Watkins' definition based on the concept of empirical content and testability seems not to grasp our intuition connected with the concept of depth in science). We confine ourselves to a merely intuitive conception of 'deeper'. The sequence (2) shows its meaning.

When we use the term 'essence' in a broad sense we may speak about the penetration into deeper and deeper layers of essence in sequence (1), too. We may say that the visible paths of planets in the sky (not mentioned there) present the surface regularities, that Copernicus disclosed the first layer of the essence of our planet system, Kepler the second layer, etc. Einstein's GR disclosed the essence of gravitation, and hence a very deep essence of motion in the Sun system and elsewhere. We say 'very deep' and not 'the deepest' because, probably, it is not the end.

Return to the sequence (2). May we speak about increasing of the truth-content here? I think we may. The later theories in this sequence use more sophisticated mathematical tools and more exact measuring methods, hence their TrC must be greater than the TrC of the previous theories.

We gave the sequence (1) as an example of progress in approximation because in this case the approximation (growing TrC) is the most characteristic trait. In the sequence (2) penetration into essence (growing D) is more characteristic. However, we may speak about the growth of D in the first and the growth of TrC in the second sequence. In general, it seems that progress in approximation (growth of TrC) and progress in penetrating the essence (growth of D) always go together.

Another aim of science postulated by Watkins (and not only by him, of course) is the unity of knowledge, in other words the scope of theories, their width. It is correlated with the depth. The deeper a theory is, the more empirical phenomena and laws it explains (in a unified way). It has then also a greater predictive power. As Watkins notices, increasing depth goes together with increasing width (pp. 219–220). I fully agree with this, I may add that this progress is accompanied with better approximation to reality. In both of our sequences, we may notice not only the growth of TrC and D but also a growth of scope, of explanatory and predictive power.

4. IS THE IDEAL OF SCIENCE BIPOLAR?

Watkins speaks about a bipolarity of the ideal of science. The first pole is the security-pole, the second the depth-pole. Thus far, we spoke about the second pole. And what about the first one?

The security-pole is the core of the Bacon-Descartes ideal of science. It required certainty of our knowledge. This ideal was continued by positivists. As Watkins rightly says, all positivists, instrumentalists, phenomenalists wage an "anti-depth war"; they are "repelled by the depth-pole because they are attracted by the security-pole" (p. 140). Watkins lays stress on the depth-pole but he preserves also the security pole, in a modified form. He replaces certain truth by possible truth. He considers high probability as an aim of science, a trait of its progress. I think, we should reject this. Undoubtedly, laws and theories must be probable, for otherwise they cannot be admitted by science. But does their probability increase in the course of the growth of science? Rather no.

The degree of certainty (the probability) is negatively correlated with generality (depth, scope, etc.). That copper is a good conductor is more certain than that each metal is good conductor. That planets orbit the Sun along ellipses (disregarding perturbations) is more certain than that all planets in the universe move so. However, science always prefers more general statements, not more certain ones. When we speak about subsequent theories in our sequences (1) and (2), it is hard to compare their degree of certainty (their probability). I do not know whether SR is more or less probable than CM, GR than SR etc. Physicists do not speak about that.

Considering rival hypotheses, physicists often say that one of them is more probable than another, using the term 'probable' in an intuitive way. But when a new theory replaces an old one, they never compare their probability. They care only for depth, unity, best approximation to reality, explanatory and predictive power, never for certainty. The ideal of contemporary science is reach but not bipolar.

Warszawa University
Warszawa

REFERENCES

Krajewski, Władysław, *Correspondence Principle and the Growth of Science* (D. Reidel Dordrecht 1977).

Popper, Karl, *Objective Knowledge* (Oxford University Press 1972).

Watkins, John, *Science and Scepticism* (Princeton University Press 1984).

JÓZEF MISIEK

PERSONAL RATIONALITY

Recently I have discovered that there are two philosophers who use the same name: Sir Karl Popper. One of them is a member of the Vienna circle, let us call him Popper$_1$. The second philosopher is a bit reluctant follower of Michael Polanyi. Let us call him Popper$_2$. It is quite easy to discern the opinions of those two philosophers. The first one is very fond of logic. The terms 'logic' or 'logical' used in all kinds of contexts mean something of the utmost scientific value to him. For instance logical analysis is the ultimate means of clarification to him. Probably this is why he never attempted a logical analysis of the term logic in this context.

Popper$_2$ is not much interested in logic, logical analysis or logical precision. He considers preoccupation with such issues as dry and unproductive. Such preoccupation results, according to him, in "scholasticism", i.e. "concentration upon minutiae (upon puzzles) and especially upon the meaning of words" which is "the ultimate cause of the dissolution of the Vienna Circle and of Logical Positivism".[1] Popper$_2$ knows very well that most important problems like the problem of research (*Forschung*) and the problem of scientific discovery are beyond the reach of logical analysis. It follows that the title of the first book published under the name of Popper was invented by both philosophers.

Popper$_1$ is afraid of psychologism. According to him the error of psychologism discredits a philosopher completely. This anxiety urges him to create epistemology without a knowing subject. Contrary to this, Popper$_2$ boldly explores epistemological problems which involve a role of a knowing subject. He calls such a discipline methodology. The main idea of this discipline is that scientific method must be considered as a theory of cognitive acts and decisions governed by the aim of scientific activity: coming closer to the truth by elimination of errors. What is more, he connects scientific activity with its rationality. "Scientific activity looks pretty much like rational activity, and since rational activity must have some aim, the attempt to describe the aim of science may be not entirely futile".[2] Such an approach could be contrasted with his *alter ego* efforts to establish epistemology upon a pure logical and even mechanical basis.

Popper's$_2$ insistence upon the role of aims or values inherent in every process of cognition and especially of scientific cognition makes him a necessary

J. Misiek (ed.), The Problem of Rationality in Science and its Philosophy, 125–130.

companion for Popper_1. The reason is that the later tacitly accepted many values, e.g. logic and precision present undisputable values to him. However, his insistence on a logical approach makes it impossible to state within his framework that he accepts such values. If not assistance of his *alter ego*, his attachment to values had to remain inexpressible.

Popper's_2 reflection upon science is quite natural for every disciple of Polanyi. "What we attempt in science is to describe and (so far as possible) explain reality".[3] He goes so far as as to acknowledge moral issues involved in scientific decisions. For instance, while describing the problem of choosing between rational and irrational attitudes, he states: "The choice between us is not simply an intellectual affair, or the matter of taste. It is a moral decision (...). Argument cannot determine such a fundamental moral decision".[4]

Popper_1 never permits such a personal approach. His ideal is objectivity. But it is very difficult to grasp what objectivity means. Sometimes he explains that it is a knowledge without a knowing subject. His favorite example of objective knowledge is knowledge published in books and journals. Objective in this context means accessible to every literate man. Sometimes he says that objective knowledge "consists of the logical content of our theories, conjectures, guesses". This definition allows us to think that everything is objective knowledge (every proposition belongs to objective knowledge) because all our theories, conjectures and guesses taken together are inconsistent. In *Unended Quest*, still another description can be found: "objective theory – a theory which is arguable, which can be exposed to rational criticism, preferably a theory which can be tested".[5]

Popper_2 is much less occupied with theories than with problems. And even when he deals with theories he makes clear that they are not testable (falsifiable) by their very nature, it rather depends upon our admission that we may reject them in the light of recalcitrant evidence. Even evidence is not such before our decision to consider it as evidence. He formulates this thesis in terms of methodological rules – which all researchers have to obey (moral obligation?). Such rules have to be adopted "as will ensure the testability of scientific statements; which is to say, their falsifiability".[6] It follows, that theories are not scientific or metaphysical *per se* but rather it depends upon our critical (=rational) or uncritical (= irrational) attitude towards them. Popper_2 is quite explicit on this topic: "A system such as classical mechanics may be 'scientific' to any degree you like; but those who uphold it dogmatically – believing perhaps, that it is their business to defend such a successful system against criticism as long as it is not conclusively disproved – are adopting the very reverse of that critical attitude which in my view is the proper one for the scientist".[7]

Thus $Popper_2$ deals with science understood as cognitive actions. His proposal amounts to stating some norms of scientific behavior; norms of rational attitude. Scientific knowledge is secondary from this point of view. The theory of knowledge or methodology is concerned with the problem of systematic connection between norms. It follows that it is $Popper_2$'s critical attitude, a kind of a moral obligation of every scientist, which makes scientific theories falsifiable. In this way $Popper_1$ profits from $Popper_2$.

$Popper_1$ is convinced that methodology makes pronouncements which are "for the most part conventions of the fairly obvious kind. Profound truths are not to be expected of methodology".[8] It fits well with Wittgenstein's opinion that "the riddle does not exist". But $Popper_2$ is quick to add a footnote that nevertheless there are such theorems of methodology which "are perhaps unexpected and not quite on the surface".[9]

As regards the method which is to be adopted in order to construct methodology, $Popper_2$ is advising us "to learn from the way in which mathematicians and scientists proceed".[10] Unfortunately, $Popper_1$ rejects such a proposal as "the naturalistic approach", which is uncritical. Instead of that he accepts a form of "conventionalism" or better pragmatism: two systems of methodological rules are to be compared as far as their usefulness is concerned. It is very close to Carnap's view that no formal system of logic or mathematics is true or false but useful or not.

So far our comparison of two philosophers yielded the following results: $Popper_1$ is a Logical Positivist who attempts to comply with the overambitious ideal of objective knowledge. This ideal is certainly derived from his fascination with formal logic. $Popper_2$ is a disciple of Michael Polanyi who prefers to deal with important problems, even if he is unable to deal with them in a strictly logical manner. The first philosopher is seriously afraid to commit the mortal *sin* of psychologism. $Popper_2$ is rather a bold philosopher but even he sometimes contracts some anxiety from his companion. $Popper_2$ is necessary to $Popper_1$ for he enables him to express his system of values. In particular, the former ensures that the notion of falsification, which is crucial for the latter, makes sense. But $Popper_1$ is also necessary to $Popper_2$ as a starting point towards more adventurous enterprises and as powerful protection against criticism by neopositivists. On the other hand, $Popper_2$ makes $Popper_1$ a fascinating writer not only for companion philosophers but for scientists also. In this sense there is unity between them.

Speaking more seriously, Popper is an example of a philosopher who attempted to free himself from the neopositivistic ideology. But, unfortunately, he lived in the same place and the same time as the main representatives of this powerful movement. Therefore he was unable to achieve intellectual sovereignty, despite his efforts. His attachment to real science was unfortunately diminished by his fascination with the neopositivistic ideal of science. The result of his efforts is impressive even if his system looks like a work written by two persons. There is nothing to laugh about this because even now we are unable to resist the tremendous pressure exerted upon us by the Viennese. Even such authorities as Kurt Gödel were intimidated from publishing philosophical papers contradicting neopositivism.

Points of affinity among Popper and neopositivists are numerous. It seems that the most pernicious among them is the conviction that "the riddle does not exist". This conviction is the cause that his Polanyian approach is not distinctive enough, so that it usually remains unnoticed. And this very conviction made him blind to riddles which are abundant in science and could clearly be seen by every unbiased observer.

One of such riddles could be found in the history of special relativity.[11] It is well known, for instance from Pais's monograph,[12] that special relativity was accepted by physicists no earlier than in late twenties. Of course, there are some notable exceptions to this statement such as Planck, Laue, or Born, but the majority of physicists were hesitant for a long time. Their reservations were connected with the fact that for a long time the experimental basis for special relativity was weak and unequivocal. What is more, it even contradicted this theory. For instance, the formula for the so called 'relativistic mass' of a particle was not confirmed until 1916, by Schäffer and others (for the electron only). Earlier experiments by Kaufman (1901, 1902, 1903, 1906) were in better agreement with the formula derived by Abraham from electron theory.

Many years after the discovery of special relativity, physicists were looking for experimental refutations of it. The most conspicuous among such attempts is the experiment performed by Dayton Miller. In 1921 he discovered a non-zero ether drift. Two years later he lectured on his discovery at the meeting of the American Physical Society. He even won a prize for his discovery.

The lack of decisive experiments was not the only reason for the reservations concerning this theory. For instance in 1907 Paul Ehrenfest published a paper where he pointed out serious conceptual problems within special relativity. Einstein answered this paper but he was unable to answer Ehrenfest's questions. He objected only to the terminology. Special relativity is not a theory – was the essence of this answer – it is a heuristic principle (Einstein's name for this theory was 'the principle of relativity').

Thus we come to a riddle: Einstein completely disregarded both experimental and conceptual objections against special relativity. One has to remember that his famous saying: "Subtle is the Lord but malicious He is not" is his answer to Miller's result. In a letter to Besso he was even more outspoken: "I have not for a moment taken Miller's result seriously".

Only two hypotheses could be adopted within the framework of $Popper_1$'s philosophy in order to explain such behavior:

1. Einstein was a lucky dogmatist.
2. Einstein had a direct line to our Lord.

Both of them are very implausible, and the first is even more implausible than the second. This is a real riddle. This is a case when we can learn from a scientist.

Before concluding the paper I would like to state the following questions. Popper's philosophy of science is to date best among influential systems of philosophy. On the other hand it is quite confusing as is evident from our distinction between $Popper_1$ and $Popper_2$. Therefore the problem is: shall we attempt a restructuring of this system? If the answer is yes, than another question comes: whose line of restructuring is to be pursued? My guess is that $Popper_2$'s philosophy is much more promising. The reason is that if the problem of rationality could not be solved it could at least be stated within his framework.

What are the main tenets of this future philosophy of science? Being Popperian in the sense of $Popper_2$ or, in other words, Polanyian, it should change focus in an essential way. First, it should pay attention not so much to knowledge as to research, and should try to discover principles which govern it. Such principles, and not simply rules, should be construed as non-mechanical but purpose-oriented. It means that they should be principles for people who are doing research, i.e. who are motivated to seek truth. Seeking truth could mean discovery of new facts or laws, elimination of some errors, or other means of perfecting the system of existing knowledge. Thus, their task is a practical one. They have to invest their time and abilities in order to make some progress. Thus, they have to know what, in the existing knowledge, is imperfect and in what way. Then they have to decide which imperfections are the most urgent and how to improve them. They have to decide which avenue of research is the most promising one. In other words, they have to have some standards of comparison between that which already is and that which should be. Other criteria would be necessary to decide whether progress in a desired direction has been achieved or not. Their decisions are personal in the sense that nonmechanical criteria are used but there is nothing subjective in them because results of their efforts speak for themselves.

Such philosophy transcends the limits of the discovery/justification distinction mainly because in every research, discovery is the main objective, including discovery of justification. On the other hand, every successful justification is only an opportunity for another discovery.

Such philosophy of science, if successfully developed, will be quite different from the dry picture of science proposed by neopositivists, the picture which serves no aim, except perhaps contemplation. The philosophy which we really need is a philosophy of action, which could serve as a guide-line for a practicing scientist.

The last statement is a warning: don't think that Popper is the only philosopher with the split personality. It is the case for all philosophers of science. Popper's case is simply an illustration of Kuhn's thesis concerning the power of paradigms. The irony is that philosophers so far ridiculed scientists for their attachment to paradigms but they forgot to look at their own. It is much easier to find paradigms in philosophy of science than in science itself.

Institute of Philosophy
Jagiellonian University
Kraków

NOTES

[1] *Unended Quest*, p. 90.

[2] 'The Aim of Science', p. 191. In: *Objective Knowledge. An Evolutionary Approach*, Oxford, Clarendon Press 1973.

[3] 'Realism', p. 222. In: David Miller (Ed.), *A Pocket Popper*, Fontana Press 1983.

[4] 'The Defence of Rationalism', p. 36. In: David Miller (Ed.), *A Pocket Popper*, Fontana Press 1983.

[5] *Unended Quest*, p. 138.

[6] 'Scientific Method', p. 136. In: David Miller (Ed.), *A Pocket Popper*, Fontana Press 1983.

[7] Ibid., p. 137.

[8] Ibid., p. 141.

[9] Ibid., p. 141.

[10] *Unended Quest*, p. 89.

[11] J. Misiek, 'Assessment of Theories', this volume. It is an improved version of 'Ocena teorii – przyczynek do problemu racjonalności nauki', (in Polish) *Zagadnienia Naukoznawstwa*, no 3–4, p. 347, 1987. See also by the same author: 'Assessment of Theories an the Problem of Rationality', in: G. Schurz and P. Weingartner (Eds), *Reports of the 11th L. Wittgenstein Symposium. Recent Developements in Epistemology and Philosophy of Science*. Hölder-Pichler-Tempsky, Vienna 1986, p. 224.

[12] A. Pais, '*Subtle is the Lord ...*', *The Science and Life of Albert Einstein*, Oxford University Press, Oxford 1982.

JÓZEF MISIEK

ASSESSMENT OF THEORIES

INTRODUCTION

The aim of this paper is to inquire into the problem of the assessment of theories. In order to be specific, I have decided to begin with the discussion of an example – historical assessment of the Special Theory of Relativity (hereafter SR). This particular example was chosen because SR has many advantages for philosophical analysis. It is far from trivial (like 'all swans are white') but from a mathematical point of view it is very simple. Moreover, SR was created in the 20th century and therefore its history is well known.

The second part of the paper is devoted to some amendments of the Popper and Lakatos methodologies in the spirit of M. Polanyi,[1] which seem necessary in view of that will be said in the first part. The relation between those two parts is the same as relation between experiment and theory in science.

I. ASSESSMENT OF SR

1. The Early History of SR[2]

In 1905 Einstein published his first paper on SR: 'Zur Elektrodynamik der bewegten Körper'. The paper appeared in Ann. Phys., *17*. The paper 'Ist die Trägheit eines Körpers von seinem Energieinhalt abhängig?' appeared in the next volume of the same journal.

In the next year Einstein published two more papers on SR. In 1907 Einstein published three papers and a long review paper of 51 pages: 'Über das Relativitätsprinzip und die aus demselben gezogenen Folgerungen', in which the last (fifth) section discusses the problem of gravitation. This paper marks the beginning of Einstein's work on general relativity; his interest in SR diminishes.

In 1908, Einstein published three papers on SR, all in cooperation with Laub. In 1909 only one paper appeared, critical remarks on Mirimanoff's paper, and in 1910 another one, semi-popular in character. It can be stated that *by 1909 Einstein's work on the consequences of SR was completed.*[3]

Some physicists readily accepted SR. In 1905 Albert Einstein, a third class

J. Misiek (ed.), The Problem of Rationality in Science and its Philosophy, 131–142.

technical expert at the Patent Office in Bern, received a letter from Prof. Planck. In 1906 he had three visitors: Max von Laue, Johan Jacob Laub, and Rudolf Ladenburg.

In the winter semester of 1905/1906, Planck lectured on SR. In 1906, his paper on SR appeared. In 1907 Laue explained the Fizeau experiment: he derived the Fizean empirical formula from SR. In the same year Ehrenfest published a short note on SR, and Einstein wrote a reply to it.

In 1911, the first monograph on SR appeared, *Das Relativitatsprinzip*, written by Max von Laue. Later monographs were written by Freundlich (1916), Born (1920), Eddington (1920), Weyl (1920), Pauli (1921).

It does not follow that SR was immediately accepted, and not without a reason. The experimental basis of SR was weak: most experiments were not precise and no important kinematic effects were confirmed. All experimental facts could also be explained by the electron theory. Therefore most physicists were not enthusiastic about SR.

Let us consider the following example. The formula for the so-called 'relativistic mass' of a particle was checked conclusively only in 1916 by Schäffer and others (for electrons only). Earlier experiments by Kaufman (1901, 1902, 1903, 1906) were in better agreement with the Abraham formula obtained in the electron theory for the model of rigid electron. But one must also remember that Lorentz obtained the relativistic formula earlier than Einstein, for the model of a contractible electron. Therefore no experiment of this kind can conclusively decide between the Lorentz theory and SR.

The same can be said concerning other experiments. For instance, the famous Michelson experiment could also be explained within the Lorentz theory by means of the contraction hypothesis. On the other hand, a positive outcome of the Michelson experiment would be an obvious falsification of SR (but not necessarily a falsification of the Lorentz theory).

Therefore, when in 1921 Dayton Miller found nonzero ether drift during his preliminary experiments performed at the Mount Wilson Observatory, many were impressed. Upon hearing those rumors Einstein commented: "Raffieniert ist der Herr Gott, aber boshaft ist er nicht". (Subtle is the Lord, but malicious He is not).[4] In a letter to Besso he wrote: "I have not for a moment taken Miller's results seriously".[5]

He expressed a similar opinion in the fall of 1919, when he received a cable which informed him that the bending of light by the sun was in agreement with the general theory of relativity. One of the students asked what he would have said if there had been no confirmation. Einstein answered: "Da Könt mir halt der Liebe Gott leid tun. Die Theorie stimmt doch". (Than I would have to pity the dear Lord. The Theory is correct anyway).[6]

One conclusion seems to follow from what was said above: although for many years SR had very weak and nonconclusive empirical evidence, Einstein, was still sure that the theory is correct. This conclusion leads to the following questions:

1. Was Einstein's attitude rational?
2. What kind of tacit methodology enabled him to adopt such an attitude?
3. Is there an agreement between Einstein's methodology and any known system of methodology, for instance the Popperian system?

The lack of crucial experiments was not the only reason for the reservations about SR. It seems that Michelson expressed this common opinion most clearly (in his 1927 book *Studies in optics*). After stating that SR "must be accorded a generous acceptance" he gave a clear relativistic explanation of the Michelson – Morley and Fizeau experiments. And next he concluded:

> The existence of an ether appears to be inconsistent with the theory. . . But without a medium how can the propagation of light waves be explained? How to explain the constancy of propagation, the fundamental assumption (at least of restricted theory) if there be no medium?[7]

Pais makes the following comment: "This is the lament not of a single individual but of an era. (. . .) In the early days it was easier to understand the mathematics of special relativity than the physics".[8]

In this context it is perhaps less striking that when in 1922 Einstein was awarded the Nobel Prize, SR and GR were not even mentioned in the Nobel Committee's statement.

It seems that there were two facts that had decisive influence on the acceptance of SR. The first of them occurred in 1919 and was of a rather psychological character: the measurement of the bending of light, which proved to be in agreement with GR. The second, and perhaps the most important fact, is connected with the discoveries in the field of nuclear physics: mass defect of the nucleus, particle and antiparticle annihilation, nuclear energy, all these confirm Einstein's celebrated theorem of the equivalence of mass and energy.

The historical developments concerning the appreciation of SR can be recapitulated in the following way. SR was readily accepted first by an elite consisting of both young and older physicists. Due to the influence of this elite, the appreciation of SR grew steadily, especially among theoreticians. Experimental physicists were more hesitant. The theory gained full acceptance in the thirties or forties.

2. Fox's Analysis or a Bit of the Contemporary History of SR

The history of the appreciation of SR does not end in the forties. One example taken from the contemporary history of this theory deserves to be described in some detail. The story begins with the paper written by J. G. Fox: 'Experimental Evidence for the Second Postulate of Special Relativity' in 1962.[9]

The Second Postulate is Einstein's Principle of the Constancy of the Velocity of Light. Fox discusses this principle *vis-á-vis* the Ritz Principle, which states that velocity of light is constant and equals c, relative to its source. The velocity of light relative to the observer is the vector sum of the velocity of light relative to the source and the velocity of the source relative to the observer. Fox's conclusion is very interesting: many experiments which were allegedly crucial for the Einstein Principle and against the Ritz Principle are not conclusive.

In order to explain this thesis I shall first present the Ritz theory and next the so called Extinction theorem of Ewald and Ossen which serves as a primary tool in Fox's argumentation.

The Ritz Theory[10]

The Ritz theory, or rather an outline of a theory, because its author died before has was able to develop his proposal, aims to describe electromagnetic phenomena in accordance with the Galilean Principle of Relativity. In order to build such a Galilean covariant theory, Ritz modifies the Maxwell-Lorentz equations.

Strictly speaking, he modifies only the nonhomogenous equations (i.e. equations containing the sources of the electromagnetic field: charge and current) in such a manner that the velocity of an electromagnetic wave equals c relative to its source.

Ritz postulates the connection of such equations, that we shall call Maxwell-Ritz equations with a picture of 'fictitious' particles which every electric charge continually emits in all directions with the velocity c (relative to the charge). The concept of the 'fictitious' particles allowed him to make a clear statement about the velocity of light in vacuum, even for a variable velocity of the source. Because the velocity of particles after emission is constant (with respect to any inertial system), their velocity relative to the observer must be equal to the vector sum of the velocity c and the velocity of the source relative to the observer at the moment of emission.

Ritz's ideas were discovered and developed independently by C. Tolman,

D. F. Comstock and O. M. Stewart. But their theories are not identical with that proposed by Ritz because all these authors introduce additional hypotheses in order to account for experiments. All such theories are called 'emission theories'.

The early discussions of Ritz's theory led to the conclusion that all first order experiments are irrelevant to it because it does not differ from SR as far as the first order effects are considered. It was also noticed that some second order experiments, e.g. the Michelson experiment, cannot decide the choice between those theories. On the other hand, it was assumed that the Michelson-type experiment with an extraterrestrial source of light would have direct bearing on the Ritz theory, because it predicts a non-zero fringe shift, contrary to the SR prediction. Such an experiment was performed by Tomaschek in 1924, with a negative result. It was considered as telling evidence against the Ritz theory.

The second case against the Ritz theory depends on de Sitter's argument: if the velocity of light equals c relative to its source, than it must have observable consequences in the case of double stars. Such stars rotate about their common center of gravity and therefore their component of velocity directed to the Earth changes periodically. This velocity adds to velocity c, and therefore light rays emitted towards the Earth change their velocity with the period of a double star. Thus, it may happen that the light emitted from one component of a double star arrives at the Earth simultaneously with the light emitted half a period later. For the Earthly observer such a star is visible at two places of its orbit. In any case the variation of the velocity of light leads to the conclusion that the observed orbit eccentricity of such star is greater than the real eccentricity. Such an effect can be measured. Astronomical data on double stars interpreted along this line lead to the conclusion that if the velocity of light depends on the velocity v of its source and equals $c + kv$ then $k < 0.002$, which provides a strong argument against the Ritz Principle.

The Extinction Theorem and Fox's Arguments

It is generally known that the velocity of light in a refracting medium, with coefficient of refraction n, is n times lower than the velocity of light *in vacuo*. It is perhaps less known that the slowing down of the light passing from vacuum to a refracting medium is not immediate: there is a finite length of the medium where the incident wave with velocity c extinguishes and a new wave, with velocity c/n is generated. The mechanism of this process is connected with the scattering of the incident wave on electrons of the medium.

The extinction theorem of Ewald and Ossen states that the incident wave extinguishes exponentially and its amplitude is reduced e ($e = 2,71\ldots$) times after a distance X equal $\lambda/2\pi(n-1)$ where λ is the wavelength of the incident wave. This distance is known as the extinction length. It follows that the incident wave, with velocity c, practically vanishes after few extinction lengths.

The extinction theorem permits easy calculation of the extinction length for any refracting medium. For instance, for glass ($n = 1.5$) the extinction length for visible light $\lambda = 6000$ Å), $X = 2 \cdot 10^{-5}$ cm. For the air in normal conditions ($n^{-1} = 2.8 \cdot 10^{-4}$ and the same wavelength range, $X = 0.04$ cm.

In rarefied gas the refraction index depends on the number of electrons N in the unit volume. In this case the following formula can be obtained:

$$X = (\lambda\, r_0\, N)\!-\!1$$

where $r_0 = e^2/mc^2 = 2.82 \cdot 10^{-13}$ cm and m is the mass and e the charge of an electron.

Now, it can easily be seen that the Tomaschek experiment is irrelevant to verification of the Ritz theory, because the velocity of light in the air is always equal to c, independently of the velocity of its source.

De Sitter's argument can also be questioned on the same grounds. First, there is evidence that many binary systems, especially those with small radii and great velocities, have a common atmosphere, which may be a cause of extinction. Interstellar matter may be another cause of this phenomenon.

Without going into details of Fox's analysis we can repeat his conclusion: "the data on binary stars do not offer any evidence against the emission theory".

It is impossible to list here all details of numerous experiments designed in order to verify the constancy of velocity of light and the Ritz theory. Fox comes to the conclusion that most of them "contain fundamental flaws which defeat their purpose". And he adds: "Our general conclusion is that there is still a good case against emission theories but that the evidence is different from and less than it has been thought to be".[11]

It is perhaps interesting to add that the unequivocal evidence against the Ritz principle until 1964 came entirely from two types of experiments: the measurement of the velocity of light transmitted through a moving glass plate, and the measurement of the velocity of decay γ rays from fast moving π^0 mesons.

As a methodological comment, I will quote Fox again:[11]

> Unless something has been overlooked, these seem to be the only pieces of experimental evidence that we have. This is surprising in view of the long history and importance of the problem. It is a curious historical fact that the recent experiments with moving sources and even the first data on time dilatation were not obtained until long after special relativity had

completely displaced the emission theory in physics.

At this moment we can raise our next problem: Is it true that the community of physicists accepted SR too early (in the forties instead of the sixties)? What is the role of experiments in the assessment of theories? What are the other (if any) factors which influence the acceptance of a theory?

Remark

It is perhaps tempting to say that the SR was accepted in the forties because at that time large indirect evidence on behalf of SR already existed. For instance, quantum electrodynamics and the Dirac equation may be cited. But one must remember that indirect evidence has never the same weight as the direct one.

For instance, Sommerfeld's explanation of the subtle structure of atomic spectra is usually considered as an indirect verification of SR (the SR prediction that the mass of an electron depends on its velocity). But the same explanation can be seen as indirect verification of the Lorentz theory.

The Dirac equation is usually given as more strong indirect evidence for SR as it describes particles with spin 1/2 in relativistic quantum mechanics. It is interesting in this context that in 1963 Dirac suggested that Lorentz's space-time symmetry is not an essential requirement – contrary to SR. He even postulated the existence of an ether.

II. METHODOLOGICAL SPECULATIONS

Popper's and Lakatos's systems of philosophy of science have many points in common. First, they both accept the thesis that the development of science is a primary philosophical problem; the structure of science is lost in oblivion. Second, both represent an anti psychological attitude and, therefore, their philosophy is concerned almost entirely with the third world (the objective mind). For them, all rational aspects of the development of science are connected with the third world; the second world seems to be a shelter for irrationality. Fourth, both connect the notion of rationality with the notion of criticism.

And here the first discrepancy appears. According to Popper, rationality is simply tantamount to criticism. According to Lakatos, rationality is a mixture of criticism and dogmatism. Dogmatism, within research programs, is a hope that theoretical progress within a program together with progress of experimental technique will, sooner or later, lead to full agreement between the program and the experiment. In Lakatos's words, Popper's theory is a theory of instant rationality. Similarly, we may call Lakatos's theory a theory of delayed rationality.

Both systems are very abstract because they do not specify the relations between objective mind on the one hand and individual minds of scientists and the 'collective mind' of a scientific community on the other. We can call it antipsychologistic hypertrophy of the objective mind over the individual or the 'sociological mind'. Because of this hypertrophy it is difficult to criticize those theories: historical, sociological and psychological facts seem to be irrelevant to them.

The aim of these speculations is not to refute Popper's and Lakatos's methodological theories as irrational (according to their own standards of rationality) or unempirical, but to amend them, to broaden their framework.

At this point an objection can be raised that there can be no connection between sociological or psychological facts and the third world methodological rules because the former are descriptive while the later are normative. One cannot investigate such normative rules by means of the empirical method. The methodological rules have to be established *a priori*.

In order to answer this objection in a few words let us consider another discipline, i.e. linguistics. The situation within methodology of science seems to be completely analogous to that of linguistics. Both disciplines have a subject of study which exists independently of them: scientific practice and language practice, respectively. Both disciplines aim at creating a normative set of rules describing the existing practice. And in both cases, practice is able to refute conjectured rules, but not *vice-versa*. Therefore, the best strategy within philosophy of science is to assume that only through the analysis of historical and other facts concerning science can we distill proper methodological rules.

The second point at which both theories must be amended is the problem of the origin of the rationality of science. Is it possible that the rationality of science is solely a product of blind chance and subsequent criticism, as is the case with mutations and the process of selection which produce fitness?

This question can be reformulated as an anthropological question: is it true that the human mind is only a mechanism generating random events? My answer to both questions is: certainly not.

Put in a positive form, the answer results in a very plausible thesis: the human mind is rational; rationality of objective science has its source in the rationality of human minds.

Yet we need to be not only rationalists but also empiricists. Therefore, I propose another thesis: men are not equal as far as their rationality is concerned. Some of them are more, and some less rational. The first thesis I shall call the principle of personal rationality, and the second of elite rationality.

But what is the rationality of the human mind? I shall adopt a tentative answer to this question. That the human mind is rational means, among other things, that it is able to gain an insight into some domain or a kind of knowledge which cannot be fully and precisely expressed in terms of the existing language before some new notions are introduced into the language or, some existing notions are made more precise. Thus the notion of insight implies the existence of knowledge which cannot be precisely expressed in terms of the existing language; it can be expressed only by means of metaphors. Such metaphorical expressions are not necessarily unclear or misty: to the contrary, they are sometimes perfectly clear and it is even difficult to recognize their metaphorical character. We shall call them, in accordance with the old tradition, 'heuristic principles'. Systems of such principles are called 'research programs'.

Thus, having begun with the problem of personal rationality, we have arrived at the problem of the development of language. According to Popper, and probably to Lakatos as well, the language of science is 'given'; its vocabulary and grammar are constant, the meaning of its terms is precise and fixed forever. The evolution of science is simply an evolution of theoretical structures within a non-evolutionary language. According to the view proposed here, language also evolves; without this evolution the progress of science is impossible.

Within this context, different degrees of personal rationality are easily exemplified: it is more difficult and more rational to invent a new notion than to understand it afterwards; it is more difficult and more rational to discover a lack of precision in a formulation or to amend it, than to understand what had been done and to use it.

According to Kazimierz Ajdukiewicz, rational knowledge fulfills two conditions:

(i) It can be formulated intersubjectively, i.e. without metaphors, similes or other imprecise ways of expression.

(ii) It can be intersubjectively checked.

As far as I know, neither Popper, nor Lakatos set forth the problem of rational knowledge; for them the only problem is that of rational method. Nevertheless, it seems that they could be sympathetic with Ajdukiewicz's definition. And here yet another amendment to their system is needed, which is perhaps a novel version of the old demarcation problem. Objective scientific knowledge must be divided into two kinds: one that fulfills both Ajdukiewicz's conditions and the other that contradicts (one of) them. The first kind is certainly more rational than the second but the latter is necessary for the development of the former. It is evident that the second kind of knowledge consists of heuristic principles and research programs. From a personal point of view, the second kind of knowledge, which is not rational in the sense of Ajdukiewicz, involves the highest standards of rationality, in the sense that it is difficult to grasp such knowledge and even more difficult to invent it. We shall call such knowledge 'elite rational'. The knowledge which fulfills Ajdukiewicz's standards we shall call 'egalitary rational'.

Now the next difference between Popper's system and my own is clear: for Popper only the egalitary rational knowledge is an object of study. It is more difficult to draw a comparison with Lakatos because his research programs which have 'heuristic power' seem to be connected only with a kind of dogmatism, and not with the violation of Ajdukiewicz's conditions. Therefore they seem to belong also to the egalitary rational knowledge.

The evolution of scientific knowledge can be described more adequately than Popper and Lakatos did. It is not only a sequence of refuted theories, as Popper claims; it is not only a sequence of abandoned research programs, as Lakatos stated. It also involves a 'diffusion' through the demarcation line delimited by Ajdukiewicz's conditions. And this last process, overlooked by most philosophers, seems to be very important when the problem of the rationality of the assessment of theories is considered. First, it seems clear that the criteria of assessment must be very different in the case of rational egalitary and rational elite knowledge. But before discussing this problem in more detail we must first consider a terminological question.

In Popper's writings there is no clear concept of theory. For instance a 'hypothesis' has for him the same meaning as a 'theory'. Things look similar in Lakatos's writings. A 'research program' seems to have the same meaning for him as a 'theory'. It seems to be connected with the fact that they both challenge the neopositivistic tradition with its emphasis on definitions and the structure of science. As a result they neglect the problem of structure.

In order to avoid confusion, I propose the following terminology. A 'hypothesis' as well as a 'law' is a proposition; the only difference between the two involves the degree of confirmation. A 'theory' is a general schema which generates models. Models are tools of prediction, explanation, etc. So, a theory never explains (or predicts) phenomena directly, there is always a mediation of models. A 'research program' is a project for constructing a theory, including suggestions as to which phenomena this (future) theory shall explain, what shall be the general character of this theory. It also contains some heuristic principles which help us to arrive at proper formulation of some laws within a theory. The hard core of a research program is usually a law and/or a model.

Also it need be stressed that some/many elements of a research program belong to the domain of elite rational knowledge. The evolution of a program is not only generation of new models by means of its heuristic principles and a theoretical or empirical knowledge external to it, but also creation of rules which enable the construction of models without heuristic principles and without external knowledge. When this aim is achieved, we can say that a program has fully evolved into a theory. In practice a name of a theory is used before all heuristic principles are expelled.

The fact that in research programs heuristic principles occur which belong to the elite rational knowledge has important consequences. The egalitary part of a program rests upon the elite one or what may be also called the metaphysical part. This part is invisible to those who are interested only in the egalitary part. Therefore their assessment of a program is different from the assessment made by those who are able to grasp also the elite part. So, it is quite normal that some people are able to see the real value of a program before other people can do so.

It follows that our view, which is an amendment of the Popper and Lakatos systems, may be contrasted with theirs in the following way: Popper's system is a doctrine of instant rationality, Lakatos's system is a philosophy of delayed rationality and our proposal involves a notion of advanced rationality.

CONCLUSION

Historical facts concerning SR seem to be in a better agreement with the amended system of Popper and Lakatos than with the original systems. The amended system makes possible not to condemn Einstein for his dogmatism avoiding also Lakatos' explanation that it is a virtue to be a dogmatist. It is a pseudo-explanation because it fails to distinguish between Einstein's brand of dogmatism and the dogmatism of his opponents.

Within our proposal the historical facts pertaining to the assessment of SR are relevant to the study of scientific method. We are unable to explain them completely but instead of explaining them away we offer a new avenue of research in philosophy of science. We not only recommend empirical method to scientists but we try also to emulate this method in philosophy. In this way, we hope to learn something more about scientific method, something less trivial than the method of trial and error or the method of dogmatic endorsing of research programs. There are many obstacles on the way, in particular the ghost of psychologism and the overambitious ideal of objective knowledge which exists beyond man-made science. In order to overcome those obstacles one has to reevaluate almost everything achieved so far in philosophy of science. The task is hard but promising.

ACKNOWLEDGMENT

I should like to express my gratitude to the Oxford Course in Philosophy of Science held in Dubrovnik in 1985 for kindly inviting me to present the first version of this paper and for providing financial help.

Institute of Philosophy
Jagiellonian University
Kraków

NOTES

[1] M. Polanyi, *Science, Faith and Society*. Oxford University Press, 1946.
[2] This part rests heavily on A. Pais: '*Subtle is the Lord. . .' The Science and the Life of Albert Einstein*. Oxford University Press, 1982.
[3] Comp. op. cit. p. 153.
[4] Ibidem, p. 113.
[5] Ibidem, p. 114.
[6] Ibidem, p. 30.
[7] Ibidem, p. 114/115.
[8] Ibidem, p. 115.
[9] J. G. Fox, 'Experimental Evidence for the Second Postulate of Special Relativity'. *Am. J. Phys. 30*, 297, 1962. See also J. D. Jackson, *Classical Electrodynamics*. John Wiley and Sons, Inc., 1975.
[10] See W. Pauli, *Theory of Relativity*, Pergamon, New York 1958.
[11] J. G. Fox, 'Evidence Against Emission Theories'. *Am. J. Phys. 33*, 1 (1965).
[12] K. Ajdukiewicz, *Problems and Theories of Philosophy*. Cambridge University Press, 1973, p 5. It seems that Ajdukiewicz's definition of rational knowledge is tacitly recognized within the whole neopositivistic tradition.

ALAN MUSGRAVE

REALISM AND IDEALISATION*

Metaphysical Objections to Scientific Realism

I

Scientific realists have often been accused of hitching their wagon to some unsavoury metaphysical view or other. Many have thought that science freed of metaphysical encumbrances would not be realist science. Duhem thought so – and so, in our own times, do Bas van Fraassen and Arthur Fine.

The accusation has some substance. Realists early and late have tied their realism too closely to the metaphysics of their day. But such ties can be loosened, and then severed, without abandoning scientific realism. The realist need not think, with the Cartesians, that the physical world is a complex of clockwork mechanisms so that the only true explanations are clockwork-mechanical explanations. The realist need not think, with the Newtonians, that the physical world consists of material corpuscles and centrally-directed forces between them so that the only true explanations are Newtonian mechanical explanations. The realist need not think, with Einstein, that the physical world is deterministic so that the only true explanations are deterministic explanations. The realist need not think, with some optimistic 'naturalised philosophers' of our own day, that current science has at last hit upon the truth so that the only true explanations are those that current theories license. Scientific realism has survived the demise of Cartesian, Newtonian, perhaps even deterministic, metaphysics. It could also survive the demise of any, most, even all, of our current theories.

Of course, realists tend to accept some theories as true (or nearly so) and tend to suppose that some 'theoretical entities' are real. You would have to a pretty pessimistic realist to think that science has come up with no truths whatever and discovered no theoretical entities whatever. Most realists are also realists about *X*'s, for some theoretical *X*'s – molecules, atoms, extinct species, genes, black holes, quarks, or whatever. But realism about *X*'s is an addition to scientific realism rather than something constitutive of it. One might abandon one's realism about *X*'s, for any theoretical *X*, without thereby abandoning one's scientific realism.

J. Misiek (ed.), The Problem of Rationality in Science and its Philosophy, 143–166.

Does this mean that scientific realism is metaphysically *neutral*? Can one be a scientific realist and adopt any metaphysical picture that you like? Does Bishop Berkeley count as a scientific realist, albeit one with rather queer views about the world?

I think not. Scientific realism, as its name suggests, is a form of realism. You cannot be a scientific realist and a metaphysical anti-realist or idealist. Bishop Berkeley does not count as a scientific realist. Being a scientific realist commits you to being a realist.

What precisely does this commitment involve? It is a commitment to what we might call 'commonsense realism' or more trendily 'folk-realism'. Folk-realism is the belief that there is a real world outside of us and largely independent of us. This world is populated with objects that do not usually depend upon our doings for their existence and which continue to exist when we are not perceiving them. The folk-realist is a realist about trees and rivers, the sun and the moon, one's mother and one's toys. As the last two examples suggest, folk-realism is a metaphysic which infants come up with quite early in their careers. The baby who looks for the rattle it has just dropped shows that it does not agree with Berkeley that 'To be is to be perceived'. The baby has to grow up and study philosophy before it will take seriously the possibility that its baby-belief is false. John Mackie speculated that humans may be innately programmed to come up with folk-realism.[1] (I have a friend who, bemused by Kantian idealism, calls folk-realism by the abusive name 'Coronation Street Realism'. So be it.)

Folk-realism, like all folk-theories, is pretty vague. We need not try to make it more precise. Nor need we try to make precise the difference between folk-realist beliefs and the further beliefs which scientific realists tend to possess. For as realists see it, there is a continuity between commonsense and science. Scientific realists accept folk-realism and think that science is in the business of describing and explaining what goes on in the real world which folk-realism postulates, refining, extending, and here and there correcting commonsense beliefs as it goes. Furthermore, explaining what goes on will involve generalisations, theories, laws of nature. And an explanation will not be adequate unless the generalisations involved in it are true. Or so orthodox scientific realism supposes.

There is an objection to all this. It is said that you cannot be a scientific realist and a folk-realist, because science realistically-interpreted shows that folk-realism is utterly false. Eddington started it with his tale of the two tables, the 'table of commonsense' and the 'table of physics', and with his question "Which table is the real table?"[2] Sellars continues it with his tale of the 'manifest' and the 'scientific' images of the world and with his insistence that the objects of the former "*do not really exist – there are no such things*".[3] (As well as Coronation Street Realism there is, one might say, Park Lane Realism. Park Lane Realists ask "Truth? Beauty? Motherhood? Hm? What do they come to in terms of quarks and gluons?")

I think that Susan Stebbing said most of what needs to be said about this objection.[4] There are not literally two tables. There is just one table, about which folk-realists have certain commonsense views, and about which scientific realists have certain scientific views. When science explains the solidity of the table in terms of repulsive forces exerted by the particles of which it is composed, science does not show that the table is not solid as commonsense supposes. Realistically-interpreted science refines and here and there corrects folk-realism, but it does not show that it is root-and-branch mistaken. (As for Park Lane Realism, if truth or beauty or motherhood cannot be cashed out in terms of quarks and gluons, then so much the worse for quarks and gluons.)

So scientific realism is not completely metaphysically neutral, since the scientific realist is also a folk-realist. But such other scientific or metaphysical doctrines as scientific realists typically propound are not in themselves constitutive of scientific realism. To make them so is to encumber scientific realism with doctrines whose demise it could well survive.

The metaphysical boot can be on the other foot too. Certain criticisms of scientific realism themselves rest upon some more or less unsavoury metaphysical thesis. We have only to recall Bishop Berkeley, convinced that material objects do not exist and that folk-realism is false, consigning physics to the rubbish-bin as far its truth is concerned and fashioning an anti-realist story about its instrumental usefulness. Bishop Berkeley convinced no one except himself and Professor Luce, so we need not dwell upon his particular example. But there are other criticisms of scientific realism which, like Berkeley's, rest upon metaphysical doctrines. Some of these are very much abroad in the land. They are going to be my topic.

First, a disclaimer. Metaphysical criticisms of scientific realism are not and have never been the most important criticisms or the most difficult criticisms to rebut. The most important criticisms and the most difficult to rebut are epistemological criticisms. I have tried to deal with some of these elsewhere. I say nothing about them here except to distinguish them from the metaphysical criticisms in which I am interested and with which they are very often confused.

II

Realists think that science seeks general truths about the world. A glance at what scientists do seems to confirm this. Indeed, scientists are inveterate generalisers. Kepler figures out that Mars goes in an ellipse and straightaway announces that all the planets go in ellipses. Mendel experiments on a few thousand pea-plants and straightaway announces 'laws of heredity' which apply to all sexually-reproducing plants and animals. Examples could be multiplied *ad nauseam.* In fact, what scientists call 'facts' are usually what philosophers would call 'general hypotheses' or 'theories'.

Now there are metaphysicians who think that the search for general truths is misguided because there are none. Popperians have been heard to mutter "All theories are false". Nancy Cartwright used to be fond of saying "There are no exceptionless generalisations"[5] These are extraordinary claims. To begin with, they are self-refuting – but no doubt we could repair things. Still, they are extraordinary claims. Let a generalisation be a statement of the form "All *A*'s are *B*'s", for any two possibly-complex predicates *A* and *B.* To say that all of these have exceptions is to say that for any two distinct predicates *A* and *B* there is in the world something of which *A* is true and *B* false. This is a metaphysical thesis. It is an unfalsifiable 'All-and-some' doctrine in John Watkins' sense. It is the metaphysic of the *generalisation-defying richness of the world.*

The world is full of riches, to be sure, many of them undreamt-of in our science or our philosophy. But by what metaphysical argument might we be convinced that is as full of riches as this? Leibniz thought that God had created the best of all possible worlds, and this meant (I am told) that He created the greatest variety of compossible beings. We could ask why at the creation God had a Texan mentality and thought 'The bigger the variety the better'. No doubt Leibniz had an answer to this. Still, Leibnizian insights into God's psyche at the creation are an extraordinary ground on which to be debunking the scientific realist's project.

The only other argument I know for the metaphysic of the generalisation-defying richness of the world is a crude form of the so-called 'pessimistic induction', which goes: "All generalisations so far proposed have been found to have exceptions. Therefore, all generalisations ever proposed will be found to have exceptions". But this argument is as awful as an argument can be: its premise is preposterous and its conclusion does not follow. It is preposterous to maintain that there is no generalisation to which no exception is known. So lunar eclipses are not always caused by the moon passing into the earth's shadow? So water is not always a compound of hydrogen and oxygen? So puerperal fever is not always brought about by micro-organisms? So limestone is not always formed by sedimentation? So the blood does not always circulate around the body? And so forth. (Here it is crucial to remember that our concerns are metaphysical not epistemic. There are well-worn epistemic arguments to the effect that no generalisation can be known with certainty to be true. But uncertainty and falsity are not, of course, the same.)

Faced with a list like this, pessimistic inductivists may try to save their premise by confining it to 'big' theories. (What is a 'big' theory? To find out, replace the little word 'big' by your favourite big word or phrase from the following list: paradigm, disciplinary matrix, scientific research programme, conceptual scheme, research-tradition, world-view, etc.) Pessimistic inductivists now owe us a distinction between big and little theories, and they had better not say that the big ones are those that have turned out to be false. Suppose (which I doubt) that 'big' theory can be defined in a non-question-begging way. Still, confining the pessimistic induction to big theories leaves little ones unscathed. And whether confined to big theories or not the pessimistic induction is invalid.

Nancy Cartwright stopped saying "There are no exceptionless generalisations", not because somebody pointed out that she was contradicting herself, but because somebody reminded her of the exceptionless generalisation "All men are mortal" (p. 46). She realised that she had been focussing too much on the equations of physics. She switched to a different thesis: "There are no exceptionless quantitative laws in physics". Now this thesis, if true, also subverts the scientific realist's enterprise, part of which is to state exceptionless quantitative laws.

To illustrate her new thesis, Cartwright quotes her optics textbook. There Snell's Law (or rather, an up-to-date version of it) is stated, beginning "At an interface between dielectric media, there is . . . " (how the Law continues need not concern us). It emerges later in the text that his Law is false for media which are not optically isotropic. A refined law is stated to take account of this, which begins "For any two media which are optically isotropic, at an interface between dielectrics there is . . ." (how the Refined Snell's Law continues need not concern us).

The transition from Snell's Law to Refined Snell's Law illustrates a common pattern. A law is stated generally which turns out not to hold generally, but instead to hold only in certain special or restricted conditions. The original law is refined by making explicit in its antecedent the conditions under which it holds. This practice of 'refining antecedents' is not, of course, confined to quantitative laws. Consider the false qualitative generalisation "All swans are white", and the restricted but true generalisation "All non-Australian swans are white".

What of the thesis that there are no exceptionless quantitative laws in physics? We seem to have been given a clear counterexample to that thesis. There are, to be sure, exceptions to Snell's Law. But Refined Snell's Law seems to be exceptionless. At least, no exception to it is given or mentioned. On the contrary, it is said to be "what is deemed true" by the optics text.

Perhaps Cartwright thinks, though she does not say it, that despite her optics text there are exceptions to the Refined Snell's Law, either exceptions already known but suppressed by the text, or exceptions yet to be discovered. Perhaps she thinks this of any qualitative law, however refined, that ever has been or will be proposed in the history of physics. This is a metaphysic of the *precision-defying fuzziness of the world:* for any two precise conditions *A* and *B*, there is something in the world of which *A* holds and *B* does not.

Or perhaps her view is slightly different again. It is not that all precise laws have exceptions, but rather that such exceptionless laws as there are are exceptionless because they are totally *inapplicable.* Refining the antecedent of a law reduces its domain of applicability. Snell's original law applied to all interfaces between media and was false. The Refined Snell's Law applies only to interfaces between optically isotropic media and is true: but it is true of few media since "*most* media are optically anisotropic" (p. 47). But suppose *no* media were perfectly isotropic. Then Refined Snell's Law would apply to no media and it would be no accident that no media would be exceptions to it. Then Refined Snell's Law would be an *ideal law.* Such laws seem to be vacuously true, through failure of their antecedents to apply to anything. It does not matter that the Refined Snell's Law is not, as Cartwright tells the tale, an ideal law. For other examples readily spring to mind, laws concerning ideal gases, perfectly elastic collisions, perfectly competititive markets, and so on.

Cartwright's stated thesis was "There are no exceptionless quantitative laws in physics". But her actual thesis might be that quantitative laws are either false or ideal, either have exceptions to them or are exceptionless because their antecedents fail to apply to anything actual. On the face of it, or a priori, refining antecedents might produce a generalisation which is:

(1) still false because its fleshed-out antecedent is true of some things of which its consequent is false;
(2) now (non-vacuously) true because its fleshed-out antecedent is true of some things which its consequent is also always true of;
(3) now vacuously true because its fleshed-out antecedent is true of nothing.

Cartwright's stated thesis regarding precise laws is that only option (1) can ever be realised. Her real thesis may be the weaker thesis that only options (1) and (3) can ever be realised. This weaker thesis rests upon a weaker precision-defying metaphysic than the earlier one. It rests upon the metaphysical thesis that for any two precise conditions *A* and *B*, however complex, *either* there is something in the world of which *A* is true and *B* is false, *or* there is nothing in the world of which *A* is true.

Neither the strong thesis nor the weak thesis seems coherent.[6] If there is a generalisation to which there are exceptions, then there is an exceptionless generalisation too. Let "All *A*'s are *B*'s" be a generalisation *G* to which $e_1, \ldots, e_n$ are the only exceptions. Then "All *A*'s except $e_1, \ldots, e_n$ are *B*'s" will be an exceptionless generalisation *G**. (Remember again that our concerns are metaphysical not epistemic: we are not here concerned with whether or how we could know that $e_1, \ldots, e_n$ are all the exceptions that there are to *G*.)

There are ways out of any refutation. One way out of this one is to deny that there could ever be only a finite number of exceptions to any generalisation. If there are any exceptions, then there are (actually or potentially?) infinitely many of them. You would need an (actually or potentially?) infinitely long *G** to accommodate them. And our generalisations must be finitely formulable. But why couldn't there just be one or two exceptions? Or why couldn't a (potentially) infinite number of exceptions all share a non-trivial feature *F* enabling us finitely to formulate the exceptionless generalisation "All *A*'s except the *F*'s are B's? (Feature *F* must not trivialise *G**: "All *A*'s except the exceptional ones, the ones that are not *B*'s, are *B*'s" won't do as a formulation of *G** since it is analytic.)

Another way out is to deny that 'exception-barring' generalisations like "All *A*'s except $e_1, \ldots, e_n$ are *B*'s" can represent genuine laws of nature. The antecedent of *G** is not a 'natural' predicate and could pick out no genuine property: statements of (putative) laws of nature must traffic in 'natural' predicates which might pick out genuine properties. But now we must distinguish true generalisations from laws and confine our scepticism to the latter. We must distinguish 'natural' predicates from 'unnatural' ones, and argue that law-statements can only contain 'natural' predicates. And we must argue further that to mention some feature *F* shared by all the *A*'s that are not *B*'s is always to create an unnatural predicate.

Both ways out of the refutation are, to say the least, problematic. Perhaps we ought in charity to take Cartwright's thesis to be something different yet again. It is not that all quantitative generalisations are false, nor is it that they are all either false or vacuously true. Rather it is that they are all only vacuously true, that only our option (3) can ever be realised. Option (2) will still be a will-o-the-wisp: there can be no non-vacuously true quantitative generalisations or laws because no quantitatively precise condition can hold of anything actual. And this might be justified *a priori* by a rather strong metaphysic of the precision-defying fuzziness of the world: for any precise condition *A* there is nothing in the world of which *A* is true.

Some think that Duhem already proposed and argued from this 'Reality is fuzzy'-metaphysic. John Worrall writes:

> Duhem argued that... [there is] an inevitable mismatch between theory and reality. He argued that reality must be taken to be as it is revealed to us through observation: fuzzy and imprecise. Theories on the other hand, are totally precise and hence spruce up or idealise reality. This is one of the reasons why infinitely many logically incompatible yet experimentally indistinguishable theories can always be given of the same range of phenomena. And it is also why theories cannot be literally true descriptions of reality:
>
> The mathematical symbol forged by theory applies to reality as armor to the body of a knight clad in iron: the more complicated the armor, the more subtle will the rigid metal seem to be; ... but no matter how numerous the fragments composing it the armor will never be exactly wedded to the human body being modelled.[7]

Duhem might have thought that reality is too fuzzy for any 'mathematical symbol' to be true of it. Like any good Roman Catholic, he acquired his metaphysics from Aristotle, and Aristotle was notoriously sceptical about the activities of the 'mathematicians'. So perhaps Duhem did attack scientific realism on metaphysical grounds derived more or less loosely from Aristotle.

Still, it is an odd thought. Duhem forever castigates the realist for encumbering and restricting physics with metaphysical doctrines. Now we have Duhem's *anti*-realism stemming from metaphysical views borrowed from Aristotle! Besides, Aristotelian metaphysics was proposed in the infancy of science and has long been outgrown by science itself—it is an odd basis on which to be debunking the realist's project!

Duhem's arguments can be read differently—or some of them can. It is not that no precise law can *be* (non-vacuously) true. Rather, it is that no precise law can be *known to be* (non-vacuously) true. Duhem has excellent epistemic arguments, derived from the necessary imprecision of all measurement, that the attribution of a precise value to any physical magnitude is irredeemably hypothetical. Quantitatively precise generalisations transcend the data in a way that qualitative generalisations never do. But again, uncertainty is not falsity, and our concern is metaphysical not epistemic.

The thesis we are considering is that no precise condition can be true of anything actual, so that the only true quantitative laws are ideal laws. Why should we accept this? Why should the values of physical magnitudes never be precise values? To be sure, many quantitative laws are ideal laws. To be sure, ideal laws pose problems for scientific realists (problems to which I shall return). What we seek is an argument that no physical magnitude can ever have a precise value – from which it will follow that the antecedent of no precise law can ever be true of anything actual.

Nancy Cartwright has an argument. It begins from a plausible enough metaphysical picture of the way the world is, a picture also derived (though she does not say so) from some remarks of Aristotle. It ends with the conclusion that the laws of physics *lie* – or rather, that they can only vacuously tell the truth.

III

Aristotle wrote of events which happen 'by chance', not in the sense that they are random or uncaused or undetermined, but rather in the sense that they occur at the confluence of two or more separate causal chains.[8] The slate is dislodged from the roof, kills the man walking below, and a verdict of 'accidental death' is returned. Yet the man had cause to be there when the slate fell, and the slate was caused to fall just as he passed beneath. The death is 'accidental' not because it is uncaused but because it results from a rare and idiosyncratic confluence or combination of causes. (Anticipating, would we try to come up with a law to 'cover' this event?)

Cartwright generalises: all natural phenomena happen 'by chance' in Aristotle's sense, all lie at the intersections of different causal domains and result from the composition of different causes. But the laws we use to explain natural phenomena are restricted to particular causal domains and concern particular causal factors. (We do not seek 'super-laws' to cover the composite phenomena with which we deal.) A law concerning a particular domain or causal factor is false if applied to phenomena involving intersecting domains and arising from the composition of different causes. So laws are all false – or rather, all are vacuously true through failure of their antecedents to apply to any real phenomenon. The 'falsity' of laws is, moreover, essential to their explanatory power: true laws could not explain anything, explanatory laws must be false. The argument in a nutshell is this:

(1) Natural phenomena are composite.
(2) Natural laws are simple.
(3) Simple laws are false of composite phenomena.

Therefore, natural laws are false of natural phenomena.

Now this argument is, I admit, very much a *reconstruction* of what Cartwright says. Here a few of the passages on which the reconstruction is based:

> One of the important tasks of a causal explanation is to show how various causes combine to produce the phenomenon under study. Theoretical laws are essential in calculating just what each cause contributes. But they canot do this if they are literally true; for they must ignore the action of laws from other theories to do the job. (12)

> When different kinds of causes compose, we want to explain what happens in the intersection of different domains. But the laws we use are designed only to tell truly what happens in each domain separately. (12)

> [There is] an extreme metaphysical possibility, in which I in fact believe. Covering law theorists tend to think . . . that there is a law to cover every case. I do not. I imagine that natural objects are much like people in societies. Their behaviour is constrained by some specific laws and by a handful of general principles . . . What happens on most occasions is dictated by no [single] law at all. (49)

> . . . the fundamental laws of physics do not describe true facts about reality. Rendered as descriptions of facts, they are false; amended to be true, they lose their fundamental explanatory force. (54)

> The fundamental laws of physics . . . do not tell us what the objects in their domains do. If we try to think of them this way, they are simply false . . . But if physics' basic, explanatory laws do not describe how things behave, what do they do? Once we have given up facticity, I do not know what to say. (55)

Enough! I think that my reconstruction does capture the central argument of Cartwright's book. But I will not argue that exegetical point. Instead I return to the argument itself.

The argument is valid – or at least, I shall grant that it is. Premise (1) – "Phenomena are composite" – is plausible enough. One could quarrel with its generality: could there not be some natural phenomena which do not arise from the 'composition of causes'? One could quarrel with it too by invoking the experimental method, which can create artificial phenomena which do not arise from the 'composition of causes'. But many experimental phenomena are also composite, involve the intersection of 'causal domains' and the operation of more than one law. Even if they were not, we would still want to explain naturally occurring phenomena, all or most of which are composite. No realist will want to concede that laws can only explain what happens in science laboratories! So let us concede premise (1), in its strong form, for the sake of the argument.

I will also be conceding, for the sake of the argument, premise (2) and the rejection, which lies behind it, of the idea of super 'covering laws' which do govern composite phenomena.

What is wrong with the argument, in my view, is premise (3). Cartwright illustrates what she means with Newton's Law of Universal Gravitation, as formulated by Richard Feynman:

> . . . two bodies exert a force between each other which varies inversely as the square of the distance between them, and varies directly as the product of their masses. (57)

Cartwright asks "Does this law truly describe how bodies behave?" and answers:

> Assuredly not. Feynman himself gives one reason why. 'Electricity also exerts forces inversely as the square of the distance, this time between charges . . .' It is not true that for *any* two bodies the force between them is given by the law of gravitation. Some bodies are charged bodies, and the force between them is not Gmm/r^2. Rather it is some resultant of this force with the electric force to which Feynman refers.
>
> For bodies which are both massive and charged, the law of universal gravitation and Coulomb's law (the law that gives the force between two charges) interact to determine the final force. But neither law by itself truly describes how the bodies behave. No charged objects will behave just as the law of universal gravitation says; and any massive objects will constitute a counterexample to Coulomb's law. These two laws are not true; worse, they are not even approximately true. (57)

Well, Newton's law and Coulomb's law may well be false, but it would take more than this farrago to show that they are. The law of gravity does not say that *the* force between any two bodies is Gmm/r^2. It says that there is *a* force between any two bodies which is Gmm/r^2. Newton himself was well aware that gravitational force is not the only force in nature. Newton would have laughed at the suggestion that his law of gravity can be refuted by experiments with charged objects such as magnets. Coulomb would have been similarly amused by the thought that his law can be refuted by experiments with uncharged but massive objects.

Newton's law gives you *the* force, the *resultant force,* between two bodies only on the assumption that no other force acts upon those bodies. Similarly with Coulomb's law. Both laws apply to all the bodies that there are and state (or purport to state) truths about them. But when we use these laws to explain the behaviour of bodies that are both massive and charged, we must proceed carefully, and correct the deliverances of gravitation theory in the light of electro-magnetic theory, or vice-versa. (In most such 'mixed applications' the corrections are quite trivial: electromagnetic forces are so much stronger than gravitational forces that in most cases the latter can simply be ignored.)

Cartwright, having convinced herself that the law of gravity is false, asks whether it can be restated so as to be true. She considers whether it will help to add "an implicit *ceteris paribus* modifier" thus:

> *If* there are no forces other than gravitational forces at work, *then* two bodies exert a force between each other which varies inversely as the square of the distance between them, and varies directly as the product of their masses. (58)

Cartwright concedes that this reformulated law is true. (This conflicts with her thesis that there are no exceptionless quantitative laws in physics, which supports the interpretation of her position given above.) She objects, however, that it is not very useful:

> *This* law can explain in only very simple, or ideal circumstances. It can account for why the force is as it is when just gravity is at work; but it is of no help for cases in which both gravity and electricity matter. Once the *ceteris paribus* modifier has been attached, the law of gravity is irrelevant to the more complex and interesting situations.
>
> This unhappy feature is a characteristic of explanatory laws. . . . There is a trade-off here between truth and explanatory power. (58–9)

To which I add that the reformulated law of gravity, as well as being pretty useless, is a pale reflection of its Newtonian forbear: Newton's law was meant to be universal, the restricted law is not.

Cartwright's reformulated law of gravity is a hopeless solution to a non-problem. Why cannot two or more true laws conspire together to explain some composite phenomenon which occurs "in the intersection of different domains"? Cartwright considers this view, calling it the 'vector addition account':

> When gravity and electricity are both at work, two forces are produced, one in accord with Coulomb's law, the other according to the law of universal gravitation. Each law is accurate. Both the gravitational and electric force are produced as described; the two forces then add together vectorially to yield the total 'resultant' force. (59)

According to Cartwright, this is just a "metaphor":

> . . . the 'component' forces are not there, in any but a metaphysical sense, to be added; and the laws that say they are there must also be given a metaphorical reading.
>
> . . . [the 'component' forces] are not real, occurent forces. In interaction a single force occurs – the force we call the 'resultant' – and this force is neither the force due to gravity nor the electric force. On the vector addition story, the gravitational and the electric force are both produced, yet neither exists. (59)

This is (metaphysical) assertion, not argument. And consider what the assertions mean. Suppose two equally powerful tug-o'-war teams each strains might and main to shift the other: the resultant force is zero and the centre of the rope does not move. Will Cartwright say that the only real force exerted here is the zero resultant force, that is, no force at all? Tell that, as they say, to the Marines, who make up one of the teams!

Perhaps a lingering positivism is at work here. Suppose you are imbued with the positivist metaphysical prejudice that the only real things are the observable or measurable things. Then the only real forces you will countenance are those that can be read off from observable motions, the so-called 'resultant' forces. The forces of which some so-called 'resultant force' is the resultant will not really exist.

In this connection John Stuart Mill made a slip. He said that a body propelled simultaneously by equal forces due North and due East "is caused to move in a given time exactly as far in both directions as the two forces would separately have carried it". As Cartwright says, this is nonsense: the body moves North-East; it moves neither to the North nor to the East; its North-East motion is not made up of two parts, a motion to the North and a motion to the East. Similarly, a body pulled equally in opposite directions stays still – it does not move three feet to the left and then three feet to the right (60). Quite so. The only real motions are to the North-East in the first case and no motion at all in the second. But to get from this thesis about real motions to a thesis about real forces, you need the positivist prejudice just mentioned. Now Cartwright explicitly disavows such positivist prejudices in favour of the observable or measurable. So I hesitate to push this explanation of what she says about 'composition of forces'. But it is the only explanation I can think of.

As for Mill, he immediately corrects his slip, saying that a body simultaneously acted upon by two forces "is left precisely where it would have arrived if it had been acted upon first by one of the two forces, and afterwards by the other". This subjunctive statement is all that Mill can have meant when he said that "one cause never, properly speaking, defeats or frustrates another; both have their full effect". Mill's Principle of the Composition of Causes is simply that "the joint effect of several causes is identical with the sum of their separate effects". (Mill has an interesting discussion of whether this principle holds universally.)

Mill's principle means that, metaphysically speaking, if any forces are to be regarded as 'unreal' or 'merely metaphorical', resultant forces have a better claim to those epithets. To call them 'resultant' is mere metaphor if what they 'result' from does not exist. In the tug-o'-war the zero resultant force is less real, one might say, than the equal and oppositely-directed forces produced by the two teams – less real in the sense that it is explained by them.

Despite her announced general theses, Cartwright eventually concedes that the 'vector addition account' can be defended and the truth of Coulomb's Law and the Law of Gravity preserved. However:

> . . . as a defence of the truth of fundamental laws, it has two important drawbacks. First, in many cases there are no *general* laws of interaction. Dynamics, with its vector addition law, is quite special in this respect. This is not to say that there are no truths about how this specific kind of cause combines with that, but rather that theories can seldom specify a procedure that works from one case to another. (63)

This harks back to an earlier point:

> We have many detailed and sophisticated theories about what happens within the various domains. But we have little theory about what happens in the intersection of domains. . . . For example, (*ceteris paribus*) adding salt to water decreases the cooking time of potatoes; taking the water to higher altitudes increases it. Refining, if we speak more carefully we might say instead, 'Adding salt to water while keeping the altitude constant decreases the cooking time; whereas increasing the altitude while keeping the saline content fixed increases it' . . . But neither of these tells what happens when we both add salt to the water and move to higher altitudes. (50)

Quite so. But this does not show that the two separate laws here are false. Nor does our ignorance about how two causes compose bespeak the falsehood of the separate causal hypotheses.

IV

Cartwright's example about the cooking time of potatoes is an interesting one. First, it shows that I was not wrong to impute to her an argument which applies as much to qualitative generalisations as to quantitative ones. For this example is of the former kind. (If it is objected that it is quantitative after all, because it involves increases and decreases in quantities such as cooking times, altitudes, and saline contents, then we can put the point differently and say that there is no clear-cut distinction to be drawn here.)

Second, this example can illustrate the way in which 'folk laws' give way to something better. The fact is that there is no such thing as *the* cooking time of potatoes: potatoes take various times to cook, depending on the potatoes. The relevant phenomenon here is actually 'the' boiling point of water. Again, it is not altitude which affects the boiling point so much as atmospheric pressure (which typically varies with altitude). The folk-law about cooking-time and altitude might give way to a law about how the boiling-point of water varies with atmospheric pressure. Similarly, the folk-law about adding salt might give way to a more precise law about how the boiling point of water varies with its saline content. Armed with such laws, we might explain (or predict) what would happen to the boiling point of water were both pressure and salinity to be changed. And the principle governing the way in which causes here compose would be simpler than vector addition – it would be *addition*. (I doubt that any scientist would venture such an explanation: this example of the composition of causes, like many others, is idiosyncratic and scientifically uninteresting.)

Third and finally, this example illustrates a point on which Cartwright is completely right. Adherents of the so-called 'covering law' model of explanation suppose that when existing laws fail to 'cover' some phenomenon by failing to mention some causally relevant factor, we must formulate a 'super law' which does 'cover' the phenomenon. Neither of the two laws in the example tells us what happens "when we both add salt to the water and move to higher altitudes", so we must formulate a 'super law' which does tell us this. Cartwright rightly insists that this is not what scientists do, nor is it something they need to do if they are to explain composite phenomena. Such 'super laws' lack generality, for they apply only to particular and often contrived instances where causes compose. What scientists do is keep their laws few and general. They deal with composite phenomena, if they deal with them at all, by employing the separate laws of each domain together with a variety of auxiliary assumptions. No physicist seeks a 'super law' to cover bodies that are both massive and electrically charged. (This is not the same as saying that no physicist seeks a super theory to explain both gravitational and electrical forces.) Physicists deal with bodies that are both massive and electrically charged by using Newton's Law and Coulomb's Law together with some auxiliary assumptions (including the law of vector addition of forces). In short, we can *explain* composite phenomena without having super laws that *cover* them. The 'covering law' model of explanation is false.

Auxiliary assumptions are the key to this. There is often an enormous gap between the phenomena we want to explain and the laws which do the explaining. Auxiliary assumptions (or what Cartwright calls 'approximation techniques') fill that gap. Cartwright makes three claims about them, the first two of which I think correct, the last of which I think incorrect.

The first claim is that auxiliary assumptions (or approximation techniques) are dictated neither by the laws being employed nor by the phenomena being explained. They really are *auxiliary* assumptions: auxiliary to the laws in that they do not follow from them; assumptions in that they do not follow from the phenomena to be explained either. If they followed from either the laws or the phenomena they would be explanatorily redundant – and they are not. Auxiliary assumptions are, in the literal and non-prejorative sense of the term, 'ad hoc' – devised to suit the case in hand.

Cartwright's second claim is that what she calls the 'generic-specific account of explanation' is wrong. According to that account:

> Phenomenological laws are what the fundamental laws *amount* to in the circumstances at hand . . . when fundamental laws explain a phenomenological law, the phenomenological law is deduced from the more fundamental in conjunction with a description of the circumstances in which the phenomenological law obtains. The deduction shows just what claims the fundamental laws make in the circumstances described. (103)

Simple-minded example: the fundamental law is "Ravens are black", the phenomenological law is "The big birds in the Tower of London are black", and the "description of the circumstances in which the phenomenological law obtains" is "The big birds in the Tower of London are ravens".

Cartwright observes that "this view is patently mistaken when one looks at real derivations in physics or engineering" (104). And she gives several detailed examples to illustrate her point. The generic-specific account of explanation is part and parcel of the 'covering law model' and must be rejected with it.

(A confusion intrudes here, however. Cartwright asserts that:

> It is never strict deduction that takes you from the fundamental equations at the beginning to the phenomenological laws at the end. Instead we require a variety of different approximations. (104)

She takes this to mean that something less than strict deduction, some mysterious non-deductive process, allows you to make this logical journey. But the mathematical arguments that she presents are as strictly deductive as any mathematical arguments are. Her point ought to have been that it is never strict deduction from the fundamental equations alone that makes the trip, that auxiliary assumptions or 'approximations' are needed as premises also.)

Cartwright's third point concerning auxiliary assumptions or approximations returns to the main theme of *How the Laws of Physics Lie.* The fundamental laws are false. The phenomenological laws we want to explain are, we suppose, true – they are the 'reality' in the case. How do we get false theory in touch with reality? We add to it 'approximations' which compensate for its errors and enable it to yield the truth:

> . . . approximations take us away from theory and each step away from theory moves closer towards the truth. (107)

> Adjustments are made where literal correctness does not matter very much in order to get the correct effects where we want them; and very often . . . *one distortion is put right by another.* That is why it often seems misleading to say that a particular aspect of a model is false to reality: given the other constraints that is just the way to restore the representation. (140; my italics)

Compensation of errors is interesting: in it two wrongs, carefully adjusted to one another, make a right! Bishop Berkeley complained that the mathematicians of his day were only able to get correct results in their calculations because they systematically made mistakes that cancelled one another out. Berkeley observed that there was nothing so scandalous in the reasonings of theologians. Cartwright thinks the scandal is endemic in the reasonings of physicists!

Now where two wrongs make a right, it is silly to single out one of them for special abuse. Cartwright harps on the fact that the laws are wrong and require 'correction' from auxiliary assumptions. You might as well harp on the fact that the auxiliary assumptions are wrong and require 'correction' from laws. Where two wrongs make a right, where "one distortion is put right by another", the credit for getting it right should be shared equally between them. (We may not always need two wrongs to make a right: sometimes a wrong and a right can turn the trick too! I ignore this logical complication.)

Cartwright's theory of compensating errors is a solution to a non-problem. She begins from the erroneous idea that any law is false because it takes only some things into account and not others. Is it then just a lucky accident when a false law yields the truth about phenomena? No, assumptions are added which 'correct' the false law, which compensate for its error and enable it to issue in the truth. But Coulomb's law is not false just because there are gravitational as well as electrical forces. It is just that when we use Coulomb's law to explain phenomena concerning bodies that are massive as well as charged, we have to make some assumption about whether gravitational forces will effect our results. Let the case be a simple one, in which we assume truly that gravitational forces are negligible. There is no sense in which this assumption compensates for the falsehood of Coulomb's law to yield the truth. We get the right answer because Coulomb's law is true of the electrical forces and because our assumption that gravitational forces are in this case negligible is also true.

There is a fourth epistemic point to be made here. Inference to the best explanation, crudely put, says that the theory which gives the best explanation of many phenomena is true. But suppose that the theory by itself explains none of these phenomena. Suppose that the theory must always be 'fleshed out' with ad hoc auxiliary assumptions or 'approximations'. There must always be a suspicion that 'compensation of errors' might be occurring. And the stronger that suspicion, the weaker is the 'inference to the best explanation'. Some of Cartwright's quantum mechanical examples raise just this suspicion. She says, for example, paralleling Berkeley once again, that typically "the same approximation applied at different points in the derivation yields two different incompatible predictions" (119). Cartwright's explanatory problems have the following flavour: "Here are my fundamental equations *A*. Here is a phenomenological law *B*. *Puzzle:* What do I need to add to *A* to get me – *logically* get me – to *B*?" This looks awfully like a Kuhnian puzzle which tests, not the truth of theory, but only the ingenuity of the mathematical physicist. If it is a Kuhnian puzzle, guaranteed to have a solution if only the physicist is clever enough to find it, then what price inference to the best explanation? At one point Cartwright says that scepticism regarding inference to the best explanation is her whole point: "I do not argue against taking the fundamental laws as true, but try only to counter the most persuasive argument in favour of doing so" (126). But then she should not have entitled her book *How the Laws of Physics Lie,* for to say that the laws of physics *lie* is to propound a metaphysical thesis not an epistemic one. (As for inference to the best explanation, I have argued elsewhere that as usually formulated it is a plain fallacy anyway.[10])

Let me sum up. I have made three points about Cartwright's metaphysical argument:

(1) The argument has nothing especially to do with qualitative laws and applies just as much to qualitative generalisations;
(2) The argument is unsound, its third premise is false, and its anti-realist conclusion need not be accepted;
(3) In developing the argument Cartwright provides the realist with a better account of law and explanation than the so-called 'covering law model'.

I now want to take up the question of ideal laws once more and to show how Cartwright provides the scientific realist with the means to solve some of the problems which such laws pose for scientific realism.

V

So far I have been concerned to resist the metaphysical thesis that all laws or generalisations must be ideal. But it must be admitted that some important laws are ideal. So far I have acquiesced in the simple-minded view that any ideal law is a vacuous truth because its antecedent fails to apply to anything.

The scientific realist should feel uncomfortable with this thought. Some ideal laws are thought be important, the fruits of decades of intricate scientific endeavour, to be taught to every science student. Vacuous truths, on the other hand, are easy to come by: if "All *A*'s are *B*'s" is vacuously true, then so are "All *A*'s are not *B*'s" and all of "All *A*'s are *C*'s" for any *C*. Suppose there are no optically isotropic media and that the Refined Snell's Law stated in Cartwright's optics text is an ideal law. The author of that text would not thank us to be told that the law stated is a mere vacuous truth on a par with countless others obtained by varying its consequent. The author will insist that these laws are not on a par with each other at all. But how can one vacuous truth be important and another not?

The difficulty is not, let us be clear, that vacuous truths are analytic truths. That there are no optically isotropic media would not render analytic a statement beginning "For any two media which are optically isotropic, at an interface between dielectrics there is . . .". Contrast an 'Analytic Snell's Law', got by defining a pair of media as 'Snellian' if Snell's Law holds of them, and then announcing "For any two Snellian media, at an interface between dielectrics there is . . .". What this comes to, when its definition is unpacked, is a triviality of the form "If *P*, then *P*". Ideal laws may be scientifically interesting – trivialities of the form "If *P*, then *P*" cannot be scientifically interesting.

But are ideal laws correctly construed as vacuous (though non-analytic) truths? Scientific realists think that scientific theories typically assert the existence of their theoretical entities. "Electrons all carry a certain elementary charge" is to be read "There are electrons and they all carry a certain elementary charge". But this will not do for ideal theories. Reading them this way will turn them all from vacuous truths into falsehoods. Scientists do not read them this way, for scientists do not believe that there are any perfect gases, perfectly elastic collisions, perfectly competitive markets, or whatever. The point of calling these 'ideal entities' is precisely that they are not thought to be real entities.

Talk of 'ideal entities' suggests a way out. Let us platonise. Let us regard ideal entities as platonic entities, denizens of Plato's World of Forms or Popper's World 3. Ideal laws may be mere vacuous truths about this world – but they can be non-vacuous and even important truths about that other world. This is hopeless. Natural science is about this world not some other world (not even the next one). Platonising is an epistemic disaster: how are we supposed to find out whether the Perfect Gas Law, and not some mischievous variant got by changing its consequent, is true of the perfect gases in Plato's or Popper's heaven? Platonising gives us no inkling how ideal laws can aid us in our explanatory dealings with real entities in this world.

Scientists do not platonise. Nor do they regard all ideal laws as vacuous truths. On the contrary, they think some true and some false. Some are thought true because they follow from theories which are thought true on independent grounds. Thus the Kinetic Theory of Gases entails that if the molecules of a gas are of negligible size and if there are no inter-molecular forces, then $PV=nRT$ (where P is pressure, V is volume, n is the number of moles of gas in the sample, R is the universal gas constant, and T is the absolute temperature). Mischievous variants of the Ideal Gas Law actually conflict with Kinetic Theory.

If such mischievous variants were (vacuous) truths, then the Kinetic Theory would be (vacuously) false. And that is obviously absurd.

This suggests a more promising idea. Laws support counterfactuals and the Laws of the Kinetic Theory of Gases are no exception. So why not construe the Ideal Gas Law, not as a universal material conditional of the form "If anything is *A,* then it is also *B*", but rather as a universal counterfactual conditional of the form "If anything were to be *A,* then it would also be *B*". And similarly for any ideal law. Thus construed, some ideal laws will be true and others false. For one thing universally agreed about the truth-conditions of counterfactuals is that they are not all vacuously true.

Alas, this is the only thing universally agreed about the truth-conditions of counterfactuals. For the rest, we have had in recent years an extraordinary proliferation of semantic theories of varying degrees of sophistication. The most famous of these is, of course, David Lewis's. To find out whether a counterfactual is true, says Lewis, you must peer into the 'closest' possible world(s) to the actual world in which its antecedent is true to see whether its consequent is also true in that world. But again, natural scientists peer only into the actual world, not into possible ones more or less remote from it.

The case of counterfactual ideal laws seems straightforward. Some theory *T* (say the Kinetic Theory of Gases), together with some contrary-to-fact assumption *A* (say the antecedent of the Ideal Gas Law), yields some conclusion *B* (say the consequent of the Ideal Gas Law). And we briefly indicate this argument with the counterfactual "If it were to be the case that *A*, then it would be the case that *B*". The counterfactual is true just in case theory *T* is true and the derivation of *B* from *A via T* is valid. Thus an adherent of *T* will declare mischievous variants of the counterfactual false because their consequents do not follow from their antecedents *via T.*

This employs an old theory of counterfactuals, first hinted at by John Stuart Mill, endorsed by Ramsey, and resurrected by Chisholm and others. According to this theory, counterfactuals are elliptical statements of logical consequence. The counterfactual "If it were the case that *A*, then it would be the case that *B*" is true just in case *A*, together with some unstated premise(s) *C* which are true, entails *B.* What the unstated or suppressed premise(s) *C* are must be 'read off' or reconstructed from the context of utterance of the counterfactual, in the same way as we do for straightforward enthymematic arguments.

I think that this general theory of counterfactuals has a lot going for it. One objection to it is that it renders the truth-values of counterfactuals indeterminate, because what the unstated premise(s) *C* are is quite indeterminate. The fact is, however, that many counterfactuals 'conversationally imply' their suppressed premises (or the key ones). How do I know this? Because factual questions can be answered in a roundabout way by counterfactuals. I illustrate.

Question: "Where does Lindsay live?" *Roundabout answer:* (1) "If Lindsay was a film-star, then a film-star would live in New York". *Direct answer* conversationally implied by (1): "Lindsay lives in New York". Hence argument indicated by (1): "Lindsay is a film-star. Lindsay lives in New York. Therefore, a film-star lives in New York". Since this argument is valid, the truth or falsehood of (1) turns entirely on whether the suppressed premise "Lindsay lives in New York" is true or false.

Question: "Do any film-stars live in New York?" *Roundabout answer:* (2) "If Lindsay was a film-star, then Lindsay would not live in New York". *Direct answer* conversationally implied by (2): "No film-star lives in New York". Hence argument indicated by (2): "Lindsay is a film-star. No film-star lives in New York. Therefore, Lindsay does not live in New York". Since this argument is valid, the truth or falsehood of (2) turns entirely on whether the suppressed premise "No film-star lives in New York" is true or false. (In fact, (2) is false.)

Question: "What nationality was Bizet?". *Roundabout answer:* (3) "If Bizet and Verdi had been compatriots, Verdi would have been French". *Direct answer* conversationally implied by (3): "Bizet was French". Hence argument indicated by (3): "Bizet and Verdi were compatriots. Bizet was French. Therefore, Verdi was French". Since this argument is valid, the truth-value of (3) turns entirely on the question of whether the suppressed premise "Bizet was French" is true or false. (In fact, (3) is true.)

Question: "How many babies has Julia had?". *Roundabout answer*: (4) "If Julia were to have another baby, she would have had six babies." *Direct answer* conversationally implied by (4): "Julia has had five babies". Hence argument indicated by (4): "Julia has another baby. Julia has had five babies. Therefore, Julia has six babies". Since this argument is valid, the truth-value of (4) turns entirely on whether its suppressed premise "Julia has had five babies" is true. (In fact, (4) is true since my wife Julia has had five babies.)

Enough of this digression. Ideal laws, we have seen, are typically supported by deeper theories. Where such theoretical support is lacking, there is no reason to think an ideal law is true. Indeed, we could scarcely formulate the Ideal Gas Law in the absence of the Kinetic Theory of Gases. A typical practice in science, mentioned earlier, was the practice of refining antecedents: a law turns out not to hold generally and is refined by making explicit in its antecedent the conditions under which it is thought to hold. This practice can abort: the refined law turn out to apply to nothing, to be a mere vacuous truth, scientifically uninteresting. But armed with a deeper theory, we may come to see that if the antecedent of our law were to hold of anything then its consequent would also hold. An ideal law is then promulgated. It is to be construed as a counterfactual. And it is true if the theory which supports it is true.

Implicit in this is an account of what the 'support' relation is in the widely accepted doctrine that laws support counterfactuals. It is not entailment: it is not that some law *L* entails a statement of the form "If it were to be the case that *A*, then it would also be the case that *B*". It is weaker than entailment: it is that *L* together with the indicative sub-clause *A* entails the indicative sub-clause *B*. Law *L* supports the counterfactual "If it were to be the case that *A*, then it would also be the case that *B*" just in case *L* and *A* entail *B*. And the true counterfactual ideal laws are those supported by true laws.

The 'elliptical inference' view of counterfactuals flies in the face of orthodoxy. For orthodoxy has it that while genuine laws support counterfactuals, accidentally-true generalisations do not. Consider an accidentally-true generalisation of the form "All *A* are *B*" and a counterfactual of the form "If *c* were an *A*, then *c* would be a *B*". The generalisation supports the counterfactual in the sense explained in the last paragraph. Moreover, on the elliptical inference view the counterfactual is true.

One can try to make these consequences palatable by interpreting "If *c* were an *A*, then *c* would be a *B*" to mean "If *c* were to be identical to one of the *A*s, then *c* would be identical to one of the *B*s". If all the coins in my pocket happen to be silver, then it is true that if Cuthbert were identical with one of those coins Cuthbert would be silver too. But standardly-interpreted, the counterfactual means "If *c* were *another A*, then *c* would be *another B*". The orthodox intuition is that this is false if "All *A* are *B*" is only accidentally true. The adherent of the elliptical inference view must, it seems, reject this intuition. (Anyone who accepted my second example of a page back, the one beginning with the question "Do any film-stars live in New York?", implicitly rejected it too; for "No film-star lives in New York" could only be accidentally true.)

One might modify the elliptical inference view by requiring, in those cases where a generalisation figures as a suppressed premise, that it state a genuine law. One would then require an independent account of how a genuine law differs from an accidentally-true generalisation. For one could hardly say that for a generalisation to support a counterfactual it must express a genuine law, and then go on to characterise laws by saying that they alone support counterfactuals. I have no independent account of law-hood to offer here.

This story about ideal laws rests upon no particular theory of laws except that laws 'support' counterfactual ideal laws in the sense explained. There are theories of law that render ideal laws problematic. David Armstrong and others think that a law is always like this: '*F*-ness necessitates *G*-ness', where *F* and *G* are genuine universals and necessitation is a primitive relation which holds between certain universals. (The relation is called 'primitive' to signify that if you do not understand it nothing can be done to help you!). Armstrong further insists that any genuine universal must have instances. It seems to follow that there are no ideal laws of nature. Still, we need to be told a tale about the ideal 'pseudo-laws' that are endemic in the sciences: how come some are true and some false? It is this tale that I am trying to tell. My tale is not quite finished.

So far we have asked how to construe, interpret, or give the truth-conditions of, ideal laws. This question, the semantic question, is only half of the problem. The other half of the problem is the pragmatic question: how can ideal laws be useful to us in our explanatory dealings with the non-ideal, with what there is? Cartwright gives us the key to the pragmatic question.

She shows that laws never explain phenomena by themselves. To get laws into explanatory touch with phenomena all sorts of auxiliary assumptions or approximations are required. This holds for ideal laws as well. Cartwright's realist theory of the use of ideal laws in explanation is as follows:

> To call a model an idealisation is to suggest that the model is a simplification of what occurs in reality, usually a simplification which omits some relevant features ... Sometimes the omitted factors make only an insignificant contribution to the effect under study. But that does not seem to be essential to idealisations... In calling something an idealisation it seems not so important that the contributions from omitted factors be small, but that they be ones for which we know how to correct. If the idealisation is to be of use, when the time comes to apply it to a real system we had better know how to add back the contributions of the factors that have been left out. In that case the use of idealisations does not seem to counter realism: either the omitted factors do not seem to matter much, or in principle we know how to treat them. (111)

Cartwright actually objects to this account in familiar ways: the auxiliary assumptions which connect the ideal with the actual are not derivable from the ideal law, nor does the ideal law entail a more complicated law which 'covers' the actual. Quite so. But these are not objections to the account she gives: it is not part of that account that ideal laws have these logical features. The realist account merely demands that an ideal law, combined with some auxiliary assumption which 'corrects for' the idealisation, can explain phenomena.

Consider a simple schematic example. Suppose we have an ideal law relating a parameter *a* to another parameter *b* which depends upon *a*: it says that if *a* were to have the value 0, then *b* would have the value *n*. Suppose we add the auxiliary assumption that *b* depends continuously on *a*. Then these two premises entail the conclusion that as *a* approaches 0, *b* approaches *n*. This factual conclusion is empirically testable. And it might be explained by the counterfactual ideal law and the continuity of nature assumption.

So ideal laws need not discomfit the realist. They are not all vacuous truths because their antecedents fail to apply to anything, nor are they all vacuous falsehoods because their antecedents existentially interpreted fail to apply to anything. They are universal counterfactuals, some true and some false, some scientifically interesting and some not. The realist project of truth-seeking in the sciences need not be defeated by them.

University of Otago

NOTES

* A paper delivered at a conference on 'The Aim and Rationality of Science' held at the Jagiellonian University, Kraków, Poland from 4–11 June 1989.

[1] Mackie writes: "we may well have an innate propensity to see things realistically . . . an infant may have some inborn reluctance to interpret what it sees as disorderly sequences, and will consequently be predisposed to take some patterns of repeated impressions as impressions of persisting things, and others as impressions of recurring processes of the same type" (*Problems from Locke*, Oxford: Clarendon Press, 1976, pp. 210–211).

[2] Sir A.S. Eddington, *The Nature of the Physical World*, Cambridge: Cambridge University Press, 1942, Introduction.

[3] W. Sellars, *Science, Perception and Reality*, London: Routledge and Kegan Paul, 1963, p. 126.

[4] In her *Philosophy and the Physicists*, London: Methuen and Company, 1937.

[5] N. Cartwright, *How the Laws of Physics Lie*, Oxford: Clarendon Press, 1983, p. 46. Further page numbers in the text refer to this book.

[6] Here I am indebted to Greg Currie.

[7] J. Worrall, 'Scientific Realism and Scientific Change', *The Philosophical Quarterly* vol. 32, 1982, pp. 201–231; cf. p. 213; the quotation is from P. Duhem, *The Aim and Structure of Physical Theory*, Princeton, NJ: Princeton University Press, 1954, p. 175.

[8] Aristotle, *Physics*, 196b32ff. Curiously, Aristotle confines the category of events which happen 'by chance' to events involving conscious agency, chance meetings at the market and the like. But a pebble being rolled down a stream may 'by chance' be broken in two by a rock dislodged from the bank of the stream.

[9] J.S. Mill, *System of Logic*, III, vi.

[10] See my 'The Ultimate Argument for Scientific Realism', in R. Nola (ed.), *Relativism and Realism in Science*, Dordrecht/Boston/London: Kluwer Academic Publishers, 1988, pp. 229–252; cf. pp. 237–239.

ZDZISŁAWA PIĄTEK

IS EVOLUTIONISM A SCIENTIFIC THEORY?

K.R. Popper (1974, 1976) states that the Darwinian theory of evolution is not a testable scientific theory, but only a metaphysical research programme which, although it is not testable, can be rationally justified, criticized and corrected. It can be a foundation for testable scientific theories. According to Popper's line of reasoning, one should remember that the word theory referring both to Darwin's *theory* and the synthetic theory of evolution should be understood metaphorically.

He believes that both the Darwinian theory of evolution and its contemporary version, i.e. the so-called synthetic theory of evolution have several shortcomings, namely:

1. They cannot be falsified;
2. They are tautological;
3. They do not make it possible to predict and, therefore, explain the evolution of various forms of living organisms;
4. Their informative content is close to nought.

Despite the above mentioned shortcomings, however, Popper is certain that the theory of evolution is invaluable, because:

1. Without this theory our knowledge could not have grown as fast as it has been doing since the time of C. Darwin;
2. Although it is metaphysical, it suggests many concrete investigations, e.g. the study of concrete mechanisms of adjustment to new, variable conditions in the environment;
3. It is the first non-theistic proposal of explanation for the phenomenon of adjustment.

In this context several questions arise: How does it happen that a metaphysical research programme burdened with so numerous shortcomings can have such significant advantages?

How does it happen that it can suggest programmes of research in nearly all domains of biology? Besides, according to many eminent biologists, the theory of evolution uniforms and gives sense to a large number of astounding facts which had been known to biologists and described by them long before the times of Darwin.

Trying to find answers to the above-mentioned questions I shall start with

J. Misiek (ed.), The Problem of Rationality in Science and its Philosophy, 167–174.

an analysis of the shortcomings of this theory. Let us consider what it means that the theory of evolution cannot be falsified. In his intellectual biography (1974), Popper discusses the following example:

Let us assume that we might discover life on Mars and find out that there are only three species of bacteria living there, which possess essential features characteristic to terrestrial organisms:

a. They possess the feature called heredity which consists in the fact that the offspring reproduces the parental organisms;
b. They are changeable, i.e. they possess the ability to undergo mutations which are hereditary;
c. Natural selection functions there.

Popper states that the presence of only three species of bacteria on Mars, or even of one species, would not falsify the theory of evolution whose fundamental thesis is the assertion: The great variety of living organisms on the Earth evolved from the few forms which came into existence in the primeval ocean.

The discovery of only three species of bacteria on Mars, and not of a large number of various forms and multitude of species would not falsify the theory of evolution because, as Popper puts it we should then say '. . .that these three species were the only forms among the many mutants, which were sufficiently well adjusted to survive.' (1974, p. 136).

The fact that the theory of evolution is not testable by the described situation issues from this, that, according to Popper, it does not predict the evolution of the varied forms, and therefore, it cannot explain this evolution. As Popper states, in the best case this theory predicts the evolution of various organisms 'under favourable conditions.' It is extremely difficult, however, to determine the favourable conditions under which the variety of forms appears.

I think that the above-cited Popper's argumentation does not consider the most important factor of the process of evolution, namely **time**. I believe that if during one billion (10^9) years the three species of bacteria living on Mars reproduced themselves unchangeably in the same proportion despite mutations and physico-chemical changes in the environment, the situation would falsify the theory of evolution both in its Darwinian version and the contemporary one called synthetic theory of evolution. The thing is that the situation on Mars described above would explicitly show that such important factor of the process of evolution as **natural selection** does not operate. Unchangeable reproduction of three species of bacteria running for a billion (10^9) years in variable physico-chemical conditions would be a miracle.

Both theories assume that if, in a sufficiently long period of time, heredity and variability in changing environmental conditions occur, then natural selection must start to operate. In order to predict what way the natural selection may go in a particular case, we would have to know all the factors determining it. If man had not appeared in the last minute of the process of evolution but were a Demon who watched the course of development of life from its very birth, then he would know the content of hereditary information of each organism, that is, he would know one of the elements which determine functioning of selection. Maybe, then it would appear that the range of changeability – i.e. the value of the second factor – is much more limited than we now think it to be.

I believe that the theory of evolution is non-testable for technical and not fundamental reasons, in the way in which metaphysical statements of the kind 'The invisible God created the visible world' are not testable either. Some biologists express the view that evolutionary theories are potentially testable. G.L. Stebbins (1977), believes that the difficulty in testing the theory of evolution lies in the fact that such experiments would have to be extremely costly and time-consuming. In order to test the prognoses about the frequency and direction of the evolutionary changes we would have to have access to sufficiently detailed data about the current evolutionary processes. R. Dawkins (1986), says that in order to falsify the theory of evolution it would suffice to unearth a skull of a mammal that would be more than half a billion (10^9) years old.

Let us now consider what the statement that the theory of evolution is tautological means. Popper justified this objection by pointing out that such basic notions of this theory like 'high value of adaptability,' 'high value of survival' or 'fitness' are tautologically defined. It is said that high value of survival is characteristic for those organisms which possess high value of fitness, and, in turn, high value of fitness characterizes those which survive, i.e. those which have high value of survival. It is not only tautological but also circular reasoning. On the basis of the theory of evolution it is said that a species living nowadays is adjusted to certain aspects of its environment because it has survived and has not got eliminated by the selection in competition with other species. The statements concerning adjustment or non-adjustment of particular organisms are usually formulated post hoc, and so – post hoc – is justified the statement concerning the origin of various species under favourable conditions, since, as Popper justly remarks, 'favourable conditions' are defined as those in which various species really occurred.

The above charge made by Popper would have been legitimate if there had been no ways of defining fitness independently of mere survival. S.J. Gould (1989) claims that the criteria of fitness independent of survival can be applied to nature and have been used consistently by evolutionists. In his opinion the charge of tautology hits only abstract mathematical models of evolution which consider evolution only as an alteration in numbers not as change in quality. As Gould says (1989, p. 95)

> My defense of Darwin is neither startling, novel, nor profound. I merely assert that Darwin was justified in analogizing natural selection with animal breeding. In artificial selection, a breeder's desire represents a 'change of environment' for a population. In this new environment, certain traits are superior a priori (they survive and spread by or our breeder's choice, but this is a *result* of their fitness, not a definition of it). In nature, Darwinian evolution is also a response to changing environments. Now, the key point: certain morphological, physiological, and behavioral traits should be superior a priori as designs for living in new environments. These traits confer fitness by an engineer's criterion of good design, not by the empirical fact of their survival and spread. It got colder before the woolly mammoth evolved its shaggy coat.

Gould's arguments show that there is no tautological connection between high value of survival and high value of fitness. In particular circumstances of random drift, for example, some organisms may survive despite the fact that their value of fitness is lower than that of the organisms that are eliminated.

The statement that the theory of evolution is of tautological character aroused opposition of biologists. It was shown in the discussion in the **American Naturalist** magazine just after the publication of the article by R.H. Peters (1976) who, like Popper, claimed that the essence of the theory of evolution resolved itself to tautological statements which could neither be falsified nor used to make predictions.

Long before Darwin's times the naturalists knew that there existed various ways of adjustment to the environment, and each of them produced different results, but they were not able to explain these observed phenomena. The theory of evolution may serve as a programme for actual research in the field of biology because its essence is not limited to tautological statements. In other words, in concrete research these statements have hardly any significance. In concrete investigations the interrelation between the adjustment of an organism to its environment, and its ability to survive in an immediate local environment and produce offsprings is empirically determined, and greatly resembles a complicated functional relation determined by unimaginably varied factors. On the side of the organism which is getting adjusted these factors may include: content of hereditary information, level of heterozygotizm, structure and synergy of all anatomical elements of organism, synergy of physiological mechanisms and processes, etc. On the side of the environment we may list: physico-chemical conditions, geographical arrangement, composition and density of individuals which constitute the ecological system, various forms of competition between the individuals of the same species, and between specimens of different species which live in the same ecological niche. Each of these factors may undergo changes contingently upon the others.

The theory of evolution makes it possible to explain and examine concrete processes of adjustment thanks to the fact that it points out the mechanisms by mediation of which these processes may take place. Within this theory interdependencies between the above-mentioned factors are examined and expressed in the form of rules or laws like Dollo's Law, Wright's Rule, Hardy-Weinberg's Law or evolutionary stable strategy of John Maynard Smith and Price, etc. Strategies of this kind, and not tautological statements, are used in empirical studies of evolutional processes. The achieved results bring us closer to better understanding of the process of evolution and enable us to correct the hypotheses adopted at the beginning of investigations. I can easily venture the opinion that since the dawn of science no other biological theory has undergone so many modifications and corrections as Darwin's theory. If we agree with K. Popper that 'there is no better synonym for 'rational' than 'critical',' then the theory of evolution is the most rational theory at the present stage of development of science. In fact, nearly all assertions of Darwin's theory except tautological statements were subject to critical analysis and got modified. These modifications were made by reference to new experiments and observations which became possible because the understanding of such basic notions as 'heredity,' 'variability,' and 'natural selection' got radically changed. It is manifested by the stages of development of this theory, and especially by its present – third stage of development, and debates concerning the basic mechanisms of the origin of species.

Let us now consider the third weak point of the theory of evolution, which is supposed to consist in this, that this theory does not make it possible to predict, and, therefore, explain the course of the process of evolution. Popper assumes that there exists symmetry between the possibility of explaining and predicting, i.e. that a scientific theory can explain as much as it can predict. The above thesis is also doubtful and not commonly accepted among biologists. Some of them, like e.g. J. Monod (1970), H. Szarski (1976) are of the opinion that a theory may explain much more than predict, because it can explain post hoc those events which it was not able to predict. It is so because the process of evolution is a unique historical process. Observing the unique process of total evolution one cannot formulate universal laws which would make it possible to predict its future course. Knowing the real course of evolutionary events, however, we can explain them on the basis of the known mechanisms of evolution and the particular configuration of factors which generated them. This is why J. Monod believes that evolution is a creative (in the strict sense) and irreversible in time process.

The question of possibilities of explanation and prediction of historical processes was analyzed by K. Popper in his book 'The Poverty of Historicism' (1956). He states that examining historical processes we cannot formulate laws which in their character would resemble e.g. physical or chemical laws. Popper believes that no universal laws of evolution exist, just as there are no universal laws of history. This is why there can be no scientific theory of historical development on the basis of which one could predict the future course of historical events, both these which refer to the history of mankind, and those which refer to the history of life. According to Popper, universal laws of history are trivial. In case of history of life, i.e. evolution, they would be laws like: Each fish will die when taken out of water. Each living organism deprived of the supply of energy will cease to function. In case of evolutionary processes, we can only distinguish general tendencies of development, but these do not resemble universal laws of Nature. I believe that Popper's fundamental thesis about methodological separateness of historical prediction and explanation is correct and refers to explanation of the process of evolution as well. This methodological separateness is a consequence of ontological peculiarity of living organisms as compared with the inanimate objects of nature. This peculiarity was extremely aptly and clear-sightedly discussed by A. Pichot (1967). He showed that neglecting of the essential role of time, i.e. historical dimension, while defining the way of functioning of living organisms led to the fundamental misunderstanding and inadmissible simplifications in the interpretation of the process of evolution.

Observing the controversies concerning the problem of predictability of the process of evolution I am of the opinion that there are two questions which should be distinguished:

1. capacity of prediction of the future course of evolution generally treated as a unique single process, and
2. capacity of prediction of the courses of particular evolutionary processes within this macroprocess.

Popper's arguments for the incapacity of prediction of the course of evolution refer to the first aspect, while, in his opinion, capacity of prediction of the course of evolutionary processes in the second aspect does not influence the capacity of prediction in the first aspect.

In the published in ***American Naturalist*** opinions regarding predictability of the theory of evolution many authors, like, e.g. G.L. Stebbins (1977) report examples concerning possibilities of prediction on the level of microevolution, and treat them as arguments for the prediction capacity of the theory of evolution neglecting the difference between the first and the second aspect.

At last, let us consider what Popper's statement that the informative content of the theory of evolution is close to nought means. Intuitively, the informative content of the theory of evolution is defined by Popper (1974) in the following way: the more a theory or a statement excludes or forbids, the more does it tell us. This intuitive idea leads to a definition of informative content, which to some people seemed absurd: 'The informative content of a theory is the set of statements which are incompatible with the theory.' (1974, p. 18) The more a theory excludes, the more information about the world it gives, and the empirical content of a theory grows with the increase of the degree of its falsifiability. The informative content of a theory corresponds with the logical content of this theory so, that each element of the logical content of the theory has, as its counterpart, an element of the informative content, i.e. negation of the statements included in the logical content. In other words, informative content of a theory includes theories and statements which are incompatible with this theory, i.e. competitive theories and statements. In turn, the theory of evolution belongs to the informative content of those competitive theories. As Popper puts it: 'There is infinity of unforeseeable nontrivial statements belonging to the informative content of any theory, and an exactly corresponding infinity of statements belonging to its logical content. We can therefore never know or understand all the implications of any theory or its full significance' (1974, p. 20). Complete understanding of a theory is an infinite task. Competitive theories and statements are gradually invented in the course of development of the main theory, and before they become invented, they are unknown. Therefore, with the passage of time theories can be understood better and better. Popper states that the informative content of the theory of evolution is close to nought, because in its general formulation this theory does not forbid anything. It does not have any more important competitive theories.

The charge that the informational content of theory of evolution is close to zero, is based on Popper's false conviction that theory of evolution has no competitive theories, that is, it does not prohibit anything. Let us see whether theory of evolution really does not prohibit anything and thus explains everything and as Popper puts it explaining everything, does not explain anything. Is it really so that there are no competitive theories to theory of evolution? I believe that such theories exist; these are: Lamarckism, creationism, preformism. Even as it is now, theory of evolution does not explain everything. It does not allow for:

a) inheritance of acquired characteristics thus excluding the passing of information from the proteins to DNA,

b) simultaneous emergence of species,

c) the process of evolution to be determined by a goal,

d) it requires the genetic units (genes) to remain separate and discrete when they are passed on to the next generation.

If the genetic units had not retained separate and discrete through generations but blended in the prototype of an individual, as Darwin believed, then the natural selection would not work.

In the above analysis I have presented Popper's reasons for questioning the scientific status of theory of evolution and attributing it with a character of metaphysical research programme. I believe that by questioning the legitimacy of Popper's charges (1–4) I have questioned the legitimacy of the claim that theory of evolution is not a scientific theory.

Jagiellonian University
Kraków

REFERENCES

Dawkins, R. (1986): *The Blind Watchmaker*, Longman Scientific Technical, Essex.

Gould, S.J. (1989): 'Darwin's Untimely Burial,' in: *Philosophy of Biology*, M. Ruse, ed. Macmillan Publishing Company, New York, pp. 93–100.

Monod, J. (1970): *Le Hasard et la Necessite, Essai sur la Philosophie Naturelle de la Biologie Moderne*, Ed. du Seuil, Paris.

Peters, R.H. (1976): 'Tautology in Evolution and Ecology,' *American Naturalist*, 110, pp. 1–12.

Pichot, A. (1987): 'The Strange Object of Biology,' *Fundamenta Scientiae*, Vol. 8, No. 1, pp. 9–30.

Popper, K.R. (1956): *Misère de l'Historicisme*, Librarie PLON, Paris.

Popper, K.R. (1974): 'Autobiography of Carl Popper', in: *The Philosophy of Carl Popper*, Vol. 1, P.A. Schilpp, ed. Open Court, La Salle, Illinois, pp. 3–181.

Popper, K.R. (1976): *Unended Quest, an Intellectual Autobiography*, Fontana/Collins, Glasgow.

Stebbins, G.L. (1977): 'In Defense of Evolution, Tautology or Theory?' *American Naturalist* 111, pp. 386–390.

Szarski, H. (1976): *Mechanizmy Ewolucji*, Ossolineum, Wrocław.

JAN WERSZOWIEC PŁAZOWSKI

RATIONALITY AND BEYOND

Die Welt ist Meine Empfindung
A. Schopenhauer

1. WHAT I HAVE LEARNED FROM THE HISTORY OF IDEOLOGY?

Man, possibly right, claimed himself to be the crown of all creatures. An impostor, however, has to pretend to be an anointed man. *Status quo* will not be sufficient. Then belief in man's distinguished position ought to be proved strongly. This necessity opens a large field for intellectuals, who may qualify to show that indeed Caesar is the best and the most honourable among the all possible emperors, either because of the thumb of the Lord, or his genealogy, his faculties, his beautiful voice, etc. Except the flatterers there are critics as well. Fine critique makes flattery more credible. Both, the flatterer and the critic, know well that upon success of one of these different methods, depends prosperity of their camarilas and themselves. A recipe for success must be obviously founded on a well-done classification. Every classification is based upon a suitable principle. If one of them is finely tuned we may have a temptation to find ourselves in the situation described by Quine in his famous anecdote about the *cyclopedic*. Let, the principle of our classification be claimed as *a priori*, necessary and objective metaphysical reason. The consequences must be imminent.

Homo Sapiens, as we have entitled ourselves, has been firmly impressed by his distinguished position – really distinguished, without any irony.

We are the unique creatures, for whom the World is by no means a pure natural fact, but rather a *sui generis* problematic situation. I have used Popper's term here, but I understand it in quite different sense. Let us notice the deceitfulness of the analogy of the path to watering-place. A river paving its way does not find itself in a problematic situation. The choice is made by the shape of the terrain, by physical laws, and finally by an accident too. The optimal solution is correlated neither to emotion, nor to values.

On the contrary, for the human being, the World is not given directly, as pure data. **To acknowledge a certain vision of the world we must first get to**

J. Misiek (ed.), The Problem of Rationality in Science and its Philosophy, 175–193.

believe in it. What we receive is only a matter that we have to form. How we have to form this matter strongly depends upon our feeling of the form, and our belief in the form of the very form. Here is, for me, the meaning of the ***problematic situation***. A problem is something what engages me totally. Reasoning can help me recognise or solve it, but the problem also, or even first of all, affects an emotion, a passion, a desire; moreover will initiate every action.

In 1907–1908 there was no empirical need for a great intellectual effort to rebuild the very foundations of the physical image of the World. General Relativity did not replace the older theory because of any pragmatic aim. Newtonian mechanics was still proving itself in many theoretical, and practical applications as well. Only the vision of the unity of physical world, describable on the background of a unique fundamental theory, and covered by a unified set of principles, was damaged in that period. Neither Newtonian mechanics, nor Maxwell's Theory of electro-magnetism could pretend to play a role of the unique fundament of physics. This situation was great dissatisfactory for most physicists. The dream about the simple theoretical order was removed far beyond the horizon. If the World does not agree with our feelings, a problem arises.

Every problem is a resentment against our distinguished position. Hence we desire a revenge, and we wish to confirm our unquestionable anointment. However, we are not ultimately certain of our position. We are afraid that the World could play us a trick and that at some moment we shall face a problem that can destroy our comfortable situation, so hardly achieved. We can create passionately and disinterestedly, remembering in our minds that we are also capable of a quite disinterested destruction. All efforts to create harmony involve a germ of its self-destruction, as if a premonition of an anarchic decay.

Therefore from ages we search for certainty, for guaranty, and for the recipe. The certainty, that our feeling of the Universe is better, more right, and more objective. The recipe, which permits us to act safely, but according to our wishes, born from our vision of reality. Finally, the guaranty, that the path on which we step will not to be a back-street.

The antinomy between the sense of our dignity and authority, and the fear that the 'problem of the World' could overwhelm us is the source of differences between philosophical and ideological views. We may perform many subtle analyses, but we know *a vista* that philosophy can guarantee us neither certainty, nor recipe. Philosophy can only propose, but potentiality of this proposition must be actualised. **The World is such, as we would believe it to be.** Ideology is philosophy, which we confess. The belief causes that the system of ideas and values gives us the feeling of certainty, guaranty and justice. We may ask here about the value of the 'objective' confirmation of a system by its 'real' successes. The very notion of success contains its evaluation. For the orthodox believer a success is only what confirms his belief. Anything to the contrary is not a failure but a temporary lack of success.

Now, one thing more, philosophy does not create **this** world. It is an inhabitant of the **'third world'**. Popperism is not my belief. Therefore the 'third world' does not seem to me to be an ultimate, glorious world in an eschatological sense. I shall speak cautiously. It would be enough to invent any philosophical system, to write it down, eventually to send it to print – for it become a part of objective knowledge, at least potentially. Now we could leave our masterpiece. A good fate will care that the work, not eaten by the mice, will await its discoverer who will want to take it from a shelf of the 'third world' and reanimate it; or, what is more probable, to breathe a spirit into it, and reveal it for the **real** world.

Ideology is quite a different matter. Belief does not exist without the believers. I do not want to incur anyone's disfavour, so I do not ask to whom world ideologies belong.

Ideology is an epiphenomenon existing and acting trough the group of confessors. It is, like other cultural 'species', not constant but submits to the mutations and evolution of itself and necessity of fitting to the changeable surroundings.

On one side, it is a relatively isolated form in the field of culture because it is possible to separate believers from infidels, at least partially. Speaking truly, the engaged nucleus draws the cloud of unbelievers and catechumens behind itself. The problem of demarcation is therefore very complicated.

On other side the *credo* is constant only literally, due to the changes in its exegesis, under the pressure of dynamic environment, moreover since it is influenced by other competitive ideologies. The very existence of 'base facts', 'objective knowledge', 'the passionless investigator', etc. is only apparent, as well as it 'engages less activity'.

I don't want to prolong my introduction, so I will hardly state my thesis. **Rationality is for me a consecrated symbol of ideology which pretends to be an official confession of science.** For better understanding of science and its regulative ideas we have, I think, to look at the history of ideologies and joined confessional groups.

2. *COUP D'OEIL* ON SOME REGULARITIES IN THE EVOLUTION OF AN IDEOLOGICAL SYSTEM

The process of growth and decline of any ideological system seems to develop in a distinct cycle. We can distinguish in it three main periods: the era of prophets, the canonical epoch, and the period of tolerance.

The era of prophets is the very beginning, and the condition for arising of the whole current. We can characterise it by a minority and opposition *status* of the confessional group. This period needs a brightness of the very idea, but its brightness is not equal to its precision. On the contrary, the metaphorical way of expressing the idea in this first epoch is attractive and causes it to act strongly upon emotions. Therefore it acquires fewer adherents, but more engaged confessors. This convinced elite proposes aims, maximised to such a degree, which could not be supported by predominant ideology, already partially corrupted, partially discredited in common opinion. The main idea in this period fulfils the function of a distinguishing symbol, granting the identity of the group and emphasising the select character of their society. Nobody awaits for immediate success of prophets. In this phase, the idea gives, foremost, belief in the righteousness of behaviour, moving the problems of guarantees and methods into the second plane. Such an embryonic period may last long and may end in the dying out of the new current. If, however one prophet has a spectacular success, along with a favourable weakness of his competitors, the history will go into the next phase.

The canonical epoch is the time of a quickly increasing influence of the ideology, which rapidly grows and begins to dominate, at least in some cultural area. This period is not homogenous, but, at first, the expansion requires an effort at codification and institutionalisation.

Codification has to preserve the main idea from dissolution in an eclecticism of various interpretations, which emerge from the great number of converts and proselytes of relatively various backgrounds. It also has to build a certain mechanism that would protect the main truths of faith against the risk of failure, and to set up prescriptions or other rules for orthodoxy. Thanks to this, new ideology can take their aims after the competitors, as heritage, and combine them with its own, without destroying the new primordial system.

The range of an ideology forces it to leave the informal structure of prophetic schools and go on to institutions. These possess their own prerogatives and division of functions and competencies, which guarantee efficiency. Moreover, a hierarchy is needed as a certain instrument of steering and controlling. The identity of the group is no longer determined simple by the common feelings. Now, the discipline of the members, their appurtenance to formal structure, and well-defined jurisdiction of everybody is required. Moreover, there emerge specified activities, social and economical constraints, and some formalised type of common behaviours or customs. The compactness of structure demands the loyalty of the members to a precisely defined orthodoxy. In this phase, ideology has to look like monolith with great perfection. This increases its attractiveness and arises confidence.

In the next phase, scratches appear on the monolith. It may have been due to: ***an internal imperfection*** of the official doctrine, for example, insufficiently consistent ideas, ***the outside causes*** with poor functioning of the preventive mechanisms, or ***too slow reactions*** for the appearance of new problems in the social and cultural environment. Therefore, inner opposition grows and a threat of schism appears. The ideological system tries to fight against it, proclaiming the need of the 'true' belief. It postulates a return to the sources and purity of the doctrine. Again the institutions arise, but now they have to watch over the consistency of ideas and observance of the rules. Yet the main idea remains unquestionable, and only human errors can cause the failures. Such activity gives us methods and criteria, authoritatively confirmed to be consistent with the main aim, leading to it in a fixed manner. The inquisition has no care for orthodoxy, rather watches for the discipline. Orthodoxy was reduced to 'orthomethodology'; the very essence was substituted by appearances.

In this period, ideology less stimulates creativity, more restrains it. Now, what we can name 'normal ideology', arises. This is characterised by 'bureaucracy of the activities'. The aims are well-known, hence only efficiency in their realisation ought to be evaluated. Really, it is impossible even to see the aim through the jungle of the 'uniquely correct rules'. Form dominates essence. Lack of real success must be covered by many merely apparent operations, and activities, which serve only the institution to endure.

It may happen that at this moment a small circle of authentic believers tries to restore the attractiveness of the ideology. If, moreover, there is someone with personal charisma, then there may begin the reformation and the ideology enters its last stage.

The era of tolerance is a swan's song of ideology. An attempt to turn the wheel of history back is justified by the thesis that while the institution falls, but the idea does not. The fertility of the previous periods is emphasised and its impact on cultural development recalled. Such an excellent idea cannot fall; what is only needed consists in a principle of tolerance. Maybe, our idea is not unique, but it has at least as many merits, as its competitors; and what is important, it is our idea. With some effort and good will, a dialogue could be proposed and a kind of complementarity accepted.

Now, however a paradox arises. A principle of conservation demands saving all that can be saved. But dialogue demands deep modification of the doctrine. Because of this game, the main idea loses its reality and becomes more and more an inaccessible ideal, and finally a pure metaphor. The rules no longer result from a belief in the aim. The very aim becomes an allegory of the set of rules. Man, however, needs faith. The ideology cannot provide it in this phase. So, it loses its attractiveness, slowly becoming an anachronism, a tradition, and, finally, also curiosity. Tolerance, like cancer, destroys ideology from the inside, being a symptom of the impotence of the idea.

Now the group of intellectuals recalls the principle: '***the fool man believes in a fool nonsense, the clever man believes in a clever nonsense***', and claims that the idea ought to be personalised. Personalism is already the collapse of ideology, and come back to a private idea without ability to gather people around a common issue. The very sense of ideology means combining the multitude into unity. The proposal of pluralism stands for the act of unconditional capitulation of ideology.

3. WHERE HAVE WE COME FROM, WHERE ARE WE GOING TO?

In a search for the meaning of term '***rationality***', I have checked my Polish dictionary (**Słownik Języka Polskiego**, M. Szymczak ed., PWN Warszawa 1984). It reads as follows:

> Rational – 1. «grounded on modern, scientific methods, giving good results; [. . .]»
> 2. «grounded on reason; [. . .], being guided by reason.»

The alteration seems to be apparently surprising. I think, however that it is closely related to a linguistic custom, based on the deeper feeling. Briefly speaking, **today rationality is not the ideal for science, on the contrary, science forms the ideal for rationality**. The general principle of regulation of scientific development has become a catchword, or a banner, under which science would like to appear to social consciousness.

Meanwhile, before our eyes, there is a very deep decay of the structures of science in progress. Many scientific branches, identifying their own aims, have come into being. Specialisation has been narrowing so much, that the process of interdisciplinary communication becomes impossible. Scientific production, carried by institutions, leads to a situation in which the elucidation of original ideas is more difficult than looking for a needle in a haystack. The common aims, or even such the common means like a unified language of science, seem to be a Utopia. The methodologists are not indeed the censors of science because nobody is affected by their poor rules. So they change themselves into ideologists trying to call back to common sense a belief in existence of some primary aim, presenting this aim as complying with factual situation.

In our time, the dispute over rationality does not belong to the canonical era as the dispute over clarity of ideas. It far more resembles the efforts of the period of tolerance, to make an idea so vague that we could save it at any price.

Let us look at it closely. Rationality as a method or regulative idea has not arisen along with the development of modern science. This stream was born at the beginning of the European culture. Confidence in the power and distinguished position of human reason was not only seen in philosophy. Many religions of our Judaic – Christian cultural circle also conceive of wisdom – ***σοφια*** and God's rationality – ***λογοσ*** as fundamental attribute of God, or even His emanation. A reasonable goal for these religions provides Man with a specific position between Heaven and the Earth. The Lord with His power of reason has filled the World with existence. Man collaborating with the Creator has been filling it with sense.

The beginning of science has related to an essential change of epistemic perspective, which our reason can illuminate. Human rationality, while insufficient for ***the theological*** cognition, brings to light a new, seemingly nearer aim – ***the cosmocentrism***. There was a kind of compromise, we may say, a behavioural postulate. Examining a cobweb, we can learn something about the spider; the 'third world' may impress the 'second world', but not *vice versa*.

However, the ***κοσμοσ*** is nothing more than the 'third world' of God, the perfect product of His reason. Therefore, the aim apparently has not changed, only the way. Still Bacon had precisely postulated the crown of science, like the natural theology.

The above change was to provide a cunning avoidance of an epistemological prohibition, arising from the position of God, simultaneously an object, and the universal subject, of cognition. Nevertheless, to outwit the Creator, we have to find other guarantees, beyond metaphysical intuition. It would account for the passionate aversion of rationalists for metaphysics, even in a sense of a product of reason.

The prophetic period of 'modern, rational' science is correlated with two eminent founders of the framework of our doctrine. Doctrine, which had to deprecate the sources of ideas, such as intuition, emotion, affection, as lying too deep in the human mind, was itself connected with the surface of consciousness, with intellect, reason, and their judgements articulated in a language. Then for internal tools of control as the feeling of truth, an external mechanism appears – *consensus*, making personal responsibility vague and throwing it onto the group. It increases an illusion of objectivity. The metaphysical background of the Cartesian *cogito* consists in the postulate that thought is the certain foundation of existence and cognition as well. Clarity and distinction of human thought forces its primary character. The *consensus* is needed only for the appropriate choice of a set of axioms. This confidence in human reason's power leads even to presuppose the possibility of reducing or rationalising our instincts and impulses.

Newton completed the thesis by a methodological rule, which forbids the use of any metaphysical intuition by a cognitive subject.

This two commandments, with the complementary Liebnizian *principio rationis sufficientis* guarantying epistemic accessibility to the structure of the World, became the set of articles of belief in the Cosmos as an entity fully describable by human reason. We can investigate every part of the structure of things, because of the isomorphism between the locality of ontology and the micro analyticity of reasoning. Things are given definitively, and independently of perceiving. Finally, cognition is objective, apart from both, the modes of decoding information and the internal structure of the decoder. Reason may receive the information, making suitable analysis and selection, and following these to a synthesis in form a linguistic representation, called a *theory*.

The creation of theory was to reflect the act of Creation. I emphasise, the act rather, then a process. God's thought, which was made flesh, should be made a Word again that we can fully behold his glory, full of grace and truth. Even nowadays, physicists speaking about the world, factually think about the set of mathematical formulas. The fate of such an isomorphism has a good cultural background, whereby the theory could be independent of the position of its discoverer, and of his prejudices, at least asymptotically. Man only discovers the objective structure of Nature, but does not create it in any sense.

Success has been spectacular. The mechanics founded on these simple rules explained many elementary phenomena. This not only strongly confirmed the power of the new ideology, but also was a beginning of fulfilment of the 'prophecy', claiming that Nature, simple and rational at its background, surely will be the same on all its levels. To approach the 'objective truth' happily and quickly, it was necessary to provide the best condition for 'objective reason' incarnated into the scientific society. Therefore autonomous and hierarchical institutions for scientific activity had be organised, warranting a functional current of information, a well-structured system of scientific investigation, the sources of economical support, the schools for adepts, etc. The scientific ideology slowly became the official ideology of 'modern', 'rational' societies, especially of the pragmatic West.

Science, however, is neither isolated, nor alone in its cultural environment. Simultaneously, there are art, religions, political systems, etc. The postulate of 'splendid isolation' was simple an unrealistic one, due to the factual metabolism of ideas. This reminds us that man, being a reasonable creature, also loves, feels, suffers, and desires. The emotions do not degrade the human nature, but constitute an integral part of it. Not everything that originates apart from reason is bad. The sciences henceforth proclaimed their ability to program perfect happiness, while the rules of social organisation would be prepared by mathematics. Really, such luck was proved as very unstable. For example, the marriage by computer programming is not always an idyll. Furthermore there are grotesque conceptions, such as the sperm bank of Nobel prize winners. There are however some others, having far more dangerous consequences as totalitarianism, pollution, etc.

The monolith revealed its cracks that appeared even in the area traditionally dominated by science. It must be once more emphasised that they were results from the activity of new ideas, not of the pressure of new facts. Both revolutionary conceptions of twentieth century physics were induced by incredibly poor empirical premises. We could hardly consider them as a cause of so deep revaluation in the sphere of ideas, especially those determining the inner harmony of Reality. Furthermore, having to accept these new theoretical visions, hot controversies appeared in science. They did not concern the matter of facts. The physicists, for better or worse, take advice from the facts, along with development of a more or less suitable language to express their theories. Their quarrels over a theoretical vision of world concern chiefly the metaphysical contents of a theory, or gently speaking, its interpretation.

The basic reason for such an engagement should be sought in that both of the new fundamental theories interpret the function of the subject in the way not quite agreeing with the rationalist orthodoxy. On one side, they introduce some epistemological constraints. On the other, they point to the necessity of active subject, already at the level of physical phenomena. Let us remark, without any evaluation, that curiosities, such as 'gnosis' from Princeton, or the anthropic principle, seem to be an expressive warning that the internal opposition increases.

This causes the system of immunity to switch on. A program of methodology, or preferably a scientific epistemology, arises. It had to serve to restore the powerfulness of the main idea by a rectification of aims and a more precise prescription of means. The postulate of rationality has tied down of self-referring. The World would be simple when recognised by reason; the world of reason would be simple when recognised by reason again. Reality has to be clear, thanks to science. Facing the difficulties in understanding the World, the Science must be also understood by itself. Such a postulate unexpectedly changes the perspective of rational cognition to that of ***anthropocentric***. It seems, to be dangerous for the old ideology. Therefore some tricks are needed to show that really such a situation does not take place.

This problem was sharply confronted with the option to rebuild rationalism on the ground of a precise methodology, with no surprise when an antipsychologism was declared. Now, let us look at it in a more detailed way.

Verificationism, in its original form, does not demand any objectivity for facts or phenomena. However, it was accepted that they constitute a solid background for our knowledge. We may consider knowledge as our knowledge in some specific sense, as the knowledge of mankind. The set of facts or phenomena, described in a language in accord with an appropriate set of rules, provided the maximum objectivity approachable by the subject. Theory was to be only a certain order or a functional catalogue of facts. It does not bring into knowledge anything that would appear either new, or subjective. About the correctness of the catalogue, the autonomous rules of logic were the watchdog, precisely dividing content from its form.

Operationism took one step farther, automating the semantic rules by some measuring procedures. Here, nevertheless, a question arises, to what extend the measure itself is embedded within the frame of the theory, that must be interpreted and confirmed.

Popper's conception was the highest formulation of the scientific objectivity. It, apparently, proves the distinguished position of science in the full area of human intellectual activity, i.e. the culture. We will dwell here for a moment. Not only for its overcoming of the significance of self-consciousness in science, but also because, for many people, the Popperian ***remedium*** seems to be a chance for, or at least an introduction to, returning to the purity of the rational ***canon*** today.

Popper's insight into the essence of rational ideology is this: unable to rationalise the Subject fully, we must to dismiss him from epistemology. Epistemology without the cognitive subject is not a metaphor. It is the only possibility for saving the rationality of science.

In result, we must try to eliminate the constraints of subjectivity on both, input and output of a knowledge producing machine. The classical Popperian criticism of the notion of a *pure fact* serves to resolve the dispute over the objectivity of facts. We may change empirical basis by convention. Moreover, the alternative: phenomenon or fact, a very unpleasant option from a viewpoint of scientific objectivity, could be replaced by some kind of relativity of facts to its theoretical framework.

The next step was a construction of the notion of theory in such a way that would exclude any danger of subjectivity from the theoretical level. The Platonic scheme was used, only translated into epistemological terms. The triad **χορα**[1] **– things – ideas** was replaced by another: **object of cognition – individual cognition – product of cognition**. In this consists the famous Popperian 'three worlds' conception. An autonomous 'third world' become the proper place for the objective theories, developing according to the objective laws.

From this position, Popper was able to begin an apparently convincing polemic with so-called psychological epistemology. He says, namely, that if civilisation felt, but the libraries were saved, then our successors could revive the knowledge, at least potentially. It will be possible if really there exists **Knowledge**, not only **Our Knowledge**. Popper seems to believe in that. In this argument I see an aftertaste of a vicious circle. Knowledge is objective because any intelligent subject can (possibly) reconstruct it, independently of the status of his civilisation. On the other hand, knowledge can be reconstructed because its theories form the objective product of cognition that is the unique rational invariant of subjectivity.

I do not want to meddle with a cheap polemic, pointing to the illusions of arguments beginning with 'if'. Human imagination is well proved to be relatively poor. Even science-fiction operates only on such fantasies that distort or extrapolate our experience. We can neither present ourselves with a situation of total civilisation destruction, nor have we full possibility to reconstruct even some little fragments of human knowledge on the basis of purely material relicts. Suffice it to see the problems of conservators trying to reconstruct only some craft or artistic technique of our own cultural circle, for example, an encaustic process of painting.

Any language, along with its informational function, fulfills an expressive and symbolic one, as well. The very meaning of a code depends not only upon its logical or formal semantic rules. For the full extraction of the meaning we need to have an intuition of a symbol by embedding it in a set of culturally determined inheritances of metaphors, allegories, signs, etc. We may believe that logic is universal, but we can hardly propose that feelings, prejudices, imaginations, and even stereotypes could be translated from one civilisation to another. We shall look at some examples. Let me recall many efforts to the decoding of hieroglyphs. However, we speak in this case about civilisation which in some sense we are heirs to. Far more serious problems arise, when we deal with a distant civilisation, for example, of South America. They are non-comparable already at a linguistic level, although it is still a civilisation of **people**, who live under the same Sun, on the same Earth, being part of the same Nature. The Popperian *third world*, the guidance of universal rationality, seems to be too poor for comprehension even the rationality of various social subjects. In comparison with the Platonic world of ideas, Popper's is extremely modest. We are able to make such a preparation only if the organism has died, but the human culture, the knowledge, and at least science is still living and developing.

This is a good moment to reflect again on the sense of scientific theory. Popper has rejected any understanding of theory as a formal ordering of empirical data. His criticism had a second edge too, omitted by the author. He acknowledged that in his *third world*, there exist objective and teleological laws of theories' development. The aim was imposed by objective truth. Such an aim can be approached by a converging sequence of partial truths if any iteration is made according to suitable rules, by elimination of errors.

I don't want repeat here the full criticism of Lakatos and Kuhn. Their remarks gave an essential contribution to the understanding of science, showing that the development of knowledge is a process rather, than an approximating sequence. It may happen due to many motives, but has two essential aspects: the discovering and the proving. I think moreover, that this process cannot be split into heuristic and formal phases, because of their mutual penetration. The heuristic perspective opens formal progress and *vice versa*. This perspective is also a selection factor at least as strong as explanatory power of the theory. Ideas can evolve, and also are evaluated with changes of culture. True enough, but I want to emphasise two questions, which are quite often omitted.

Firstly, Popper's remark unfortunately of small results on construction of his own system: I think about the thesis that an intimate analogy exists between natural evolution and scientific development. Thanks to this analogy, we may suppose, that the latter has two essential characteristics:

1. It is **deterministic** only **between** the appearances of **mutations**. The very moment of mutation introduces an element of chance to the system, just as a measurement in quantum mechanics. This is the real singularity of the Popper's 'third world', beyond which we cannot pass due to any internal rules of the logic of scientific discovery.
2. It seems to be **teleonomic, but not teleological**. Natural evolution does

not realise any aims, fixed in advance, e.g. creation of human beings. Scientific evolution seems to go similarly. That, what we want to see as an intentional action is, in fact, only a set of the adaptive measures. We must emphasise, adaptive not only to the cultural, but also to the natural environment, so it does not imply any kind of relativism. As it is, natural species also discover some truths by adaptation, but not regarding an aim-in-view. To learn about truth is only one mean of survival. The very notion of the aim comes from beyond every logic, beyond the 'logic' of scientific discovery too. The process of development of a scientific theory in any case is not similar to an analytic convergence. It rather resembles a game with a tree and its branches, in which a challenge to strategy depends on a variety of internal and external factors, and last but not least on the history. For an evolutionary description, the environment and the manifold of competing cultural species as well, as the sources of mutation, are missing in Popper's 'third world'.

The next aspect, important but neglected by Popper, consists in the role played by interpretation in the emerging and developing of theory. One may protest: why emerging? An interpretation is so strong a factor of heuresis, especially in the initial phase, that for me it is high time to abandon the stereotype that sees it as a secondary matter. The interpretation, above all, contains an information about the history and is a kind of a program resident in the operative memory of culture, constituting a proper part of a 'genotype' of scientific theory, we might say.

The same, or a bit more prolonged argument of Popper's critique of 'pure fact', can serve us as good ground for thesis that **a pure scientific theory without any metaphysical embedding does not exist**. Just as a pure empirical fact without any theoretical interpretation does not exist. Above thesis is rarely questioned, even by the most convinced positivists. However, they would like to see the metaphysics as an expression of the weakness of the human mind, unable to carry through a pure rational phase in practice. Therefore it is recognised as an external and strange element for science. The paraphrase of St. Matthew, which is imputed to Eddington, says: *'the flesh of facts indeed is willing, but the spirit is weak and must follow an interpretation'*. I think however that reality is far less trivial, and the metaphysical interpretation belongs to the ***proper, internal*** equipment of theory.

Speaking more clearly, the terms of theory have no sense neither by themselves, nor by linguistic context, nor by operational rules of semantics. Theory is embedded in natural language and we cannot to comprehend it without some relations to a metaphysical viewpoint. Especially, the fundamental theoretical notions, as *existence*, *substance*, *order*, *symmetry*, *unity*, etc., have a strong metaphysical connotation. The sources of the metaphors needed we can find in culture at large, but their main source lies in philosophy, principally in metaphysical systems. The philosophical commitments of theory play a more active role than aiding the demonstration of meaning. While not developing this motif further, I want to point out some moments in which metaphysic acts as an important factor for stimulation or selection:

(i) Regulating the internal structure of theory, giving the frame of the conception of ***order and harmony***. In practice, it introduces to the theory such concepts as a mode of existence, a structure of ontic levels, a correlation between these, etc.
(ii) Determining the ***system of values*** pertaining to the theory, for example, inner perfection, beauty, simplicity.
(iii) Laying down the ***heuristic plan***, e.g. geometrisation, unification, searching for global and/or local symmetry.
(iv) Serving as an element of ***accommodation to the domain***, as with the problem of Quantum Cosmology.

Therefore, metaphysics is not an anticipation, nor a horizon of theory, but a ghost that animates its scheme. Theory without fact is empty, without metaphysic is dead.

We can draw an evident analogy. Science, like art, was born from metaphysical experience of the problem of the World by the creative subject. The analogy is instructive. It seems doubtful that among aestheticians somebody will appear, who will claim that music might exist without any listener, in notes only. Apart from trivial arguments that discoverers of these notes might not possess the faculty of hearing, it is clear that aesthetic experience is so deeply rooted in the cultural context, in a metaphysical conception of harmony, in the set of typical human emotions, passions, and desires, that nobody can seriously propose any theory of art without a sensitive subject, being the creator, and simultaneously the receiver, as well.

Man jealously protects this part of his culture, which, like art, is deeply related to his own condition. Only science is seen as a proper home for objectivity. In science, we think, the seventh seal is sufficiently hidden, so we can feel safe from violation of our intimacy. Art, however, and quite similarly science, is our human vision of the World, burdened by our imagination, our evaluation and our mode of experience. They differ only by the way of expression.

Popper, aspiring to a depsychogising of epistemology, discovered, what we may call a '**rational *ego***' of the social subject. It is paradox that he claimed it to be the unique factor of a general interpersonal consciousness. His is one of those many philosophical minds that yield to the illusion. He believes that rationalisation of human activity could clarify certain rules, hitherto obscured by the burden of individual subjectivity – the mystical domain of affections and passions. The history of philosophy knows many examples of such belief. From Descartes and De Spinoza to formal methodology we are still waiting for a miracle, that permits us to find an algorithm for creativity, the most marvellous part of human intellect. Until now, this belief has been fruitless. Instead of rules for creation we have always obtained rather some censorial prescriptions, far from satisfying. Systematics, even perfect, is only systematics, and it is far from being a genetic engineering.

Having to discover the mechanism of the development of knowledge we must reconcile the fact that creation comes out of the vague, 'irrational' element of feelings embedded in subconsciousness. In the process of creation it changes, we may say, it becomes 'more rational', but this is only the surface of the whole process. Just as we cannot explain elasticity without solid state physics, nor construct elementary particles without the vacuum, we also cannot avoid to descend to the hell of subconscious, if we want to learn something more about the growth of knowledge.

Polanyi pursued this thought. He revealed for the philosophy of science the large domain of instincts, likings, habits, and prejudices, taking its part in the process of scientific development. He went even a step further, claiming these habits or abilities to be transferable in some sense, at least within the framework of a school or the surrounding. Therefore he claims them to comprise an element of knowledge.

Tacit knowledge is nothing more than attributing subconsciousness to the social subject, or if you prefer, discovering it. Polanyi supplements the Popperian over-subjective[2] '***ego***' with the '***id***'. For the philosophy of science, this step was as important and risky as Freud's conception was for psychology. Postulating some kind of deeper mechanism with a vague definition of its content, it leads to the tendency to push everything inconvenient, difficult, and doubtful down into this 'deep structure'.

With this temptation come an unfounded and superfluous enlarging of the domain of meaning of the term '*rationality*', based upon the concept that instinct for truth, the rational conscience, is located within *tacit knowledge* If all that science is doing has to be rational, without regard to the leading rules, than we must use various pure verbal resources. In that case, however, the thesis on rationality of science becomes a simple tautology.

It seems that reduction of all extra linguistic motives to *tacit knowledge* may enlarge the field of flexibility, but still science is considered as an object isolated from its cultural environment. Such an organism must develop by an approximately deterministic, internal mechanism. It is not a realistic view.

The conception of Kuhnian paradigm does not solve this problem, because of neglect of the continuity, and the attempt to find a solution in a purely behavioural manner, but not only.

The fundamental deficiency of the Popper-Polanyi scheme seems to be evident. Reason operates between the instincts and the belief. 'Nothing is so forth impressed or deeper joined with our soul, as the idea or feeling of order and symmetry'.[3] Instinctively we aspire to order, we express it by reason, but the source of the idea of order does not lie in an instinct, but in metaphysical intuition. The need for harmony may be articulated, but the conviction about the shape of harmony is taken from metaphysical philosophy. There are many possibilities, but, during the fully conscious creative process, we are adherents to one and not another conception of the order of the being.

It is enough to look at the disputes over atomism, at the foundations of quantum mechanics or at the development of the theory of symmetry in physics, to notice that the greatest creators of science do not follow the subconsciously formed visions only. On the contrary, they carried with full consciousness to adopt certain metaphysical convictions, as a more promising plan for their investigations. Let us emphasise again: the metaphysical disputes are not an indifferent supplement to science. They form a background of creativity, introducing the principles of order in accord with which the reason will develop its reasoning. Therefore metaphysics is neither part of the '***ego***' nor of the '***id***' of the scientific consciousness. It belongs to the quite another level. Continuing our psychological analogy, we may call them the '***superego***' – the culturally determined level of mind. Unlike Kuhn, I should consider this level as a proper part of the whole internal structure of the subject who creates science.

Now we have recovered the full structure of individual mind in the social subject. The creative social subject needs to have:

- an intersubjective, rational '***ego***', represented by the Popperian 'third world';
- a domain of subjectivity – '***tacit knowledge***';
- the culturally formed '***super-consciousness***' containing the manifold of metaphysical planes, which could be made concrete in theories.

The cycle closes. The escape from the real subject ends in necessity of providing the abstract social subject with the same structure.

4. THE COGNITIVE SUBJECT WITHOUT EPISTEMOLOGY

When Albert Einstein was criticised for inconsistency in his methodological approach, he not only did not try to defend himself, but conversely made a virtue from a sin. He responded:

> The scientist [. . .] accepts gracefully the epistemological conceptual analysis but **the external conditions**[4] [. . .] do not permit him to let himself be too much restricted in the construction of his conceptual world by the adherence to an epistemological system. He therefore must appear to the systematic epistemologist as a type of unscrupulous opportunist: he appears as realist, as he seeks to describe a world independent of acts of perception, as idealist insofar as he looks upon the concepts and theories as the free inventions of human spirit (not logically derivable from what is empirically given), as positivist insofar as he considers his concepts and theories justified only to the extent to which they furnish a logical representation of relations among sensory experiences. He may even appear as Platonist or Pythagorean insofar as he considers the viewpoint of logical simplicity as an indispensable and effective tool of his research.

Hitherto we have tried to show that the pure rationality as an ideal for science is an illusion.

On the one hand, it is neither an objective attribute of knowledge, nor any autotelic value, but derives from some special set of metaphysical commitments and presuppositions. The conception of order, which dominates a given epoch of scientific development imposes a consistent system of reasons. Rationality is not even a regulative idea for science because methodology tends to extent this notion so far that it could comprehend whatever has emerged in science. Science on the contrary holds before itself its own aims, without any respect for methodology. Any effort to automate the process of creation without regard to the subject, seems to be a failure. Subjectivity thrown out by the one door comes back through another.

We must, therefore, become reconciled with this 'unpleasant' fact that knowledge is ***our human*** knowledge. It contains our human vision of the World, and is burdened by our human subjective structure. It does not mean that we must literally repeat after Schopenhauer: 'the World is our feeling'. Nevertheless, I think, that our feeling takes to substantiate one and not another image of the World. Thus, our feeling gets the rule of selection one from the all possible visions of the World. It does not mean a renunciation of truth, but only of an illusion that some crude idealisation may give us a simple recipe for obtaining the truth. Rationality in such a sense is only ideology, which promises an easy, and so to say, too mechanical approach to the truth.

On the other hand, for the rationality could explain anything, it would be neither a definition nor a synonym of science, or more precisely, of successful science. If we take our aim to be such a reconstruction of the notion of rationality that it could reach recent status of scientific development, then philosophy, instead of being fertile, will be doomed to stay in the lobby and wait to join the winning competitor. Unfortunately, the competitiveness we can evaluate only *ex post*, when success became a fact.

The power of philosophy arises from the circumstances, that philosophical thinking cannot provide any ready prescriptions but only examines a large set of potential strategies. The ideology, on the contrary, claims to have a unique perfect strategy. When ideology suffers from the consequences of chance, philosophy continues to be a source of suggestions. Ideology does not investigate various systems of norms and values but imposes its own, as only identifiable and admissible system. Rationalism, alike other ideology, promises to believer the happiness of luck of responsibility for his achievements. The loyalty to the methodology guarantees the full success. The ideological norms appeal to the egoism, for example 'do as you would be done'. Similarly, rational is all what I, and what is even more important my chief, believe in. It simply contradicts the reason, and we have to reject either reason or the belief.

From its beginnings, the philosophical attitude has been dialectical. This manner of approaching truth has tried to take for account as large as possible a range of convictions, keeping in mind that it is still only a part of ***human*** convictions. Unlike the system of deduction, dialectic works in analogy to the equivalence principle in physics. There is no objective distinguished frame of ideas and values, but we always may look for an invariance. We also may investigate which singularities are real ones, and which derive from the accepted framework. We may further ask which frames are more convenient, more comprehensive, and whether the structure of the object allows for a description by a unique framework. Such a procedure seems to be more close to the spirit of real science, rather than insisting on one, accidental framework, claiming it to be uniquely distinguished by 'resting in the ether'.

It seems to me that the more reasonable epistemological postulate would be the full subject without epistemology, rather then epistemology without a subject. Precisely, without epistemology conceived as a dogma, because this, in essence, is only an ideology.

Maybe, what I am saying here will sound paradoxical, but for me the Feyerabend claims are no anarchism. On the contrary they sound like exhortation to tolerance. He tries to achieve so vague an ideal of rationality that it could be agreed with any other regulative idea of science. I think that the very fact so excites the orthodox philosophers, seeing clearly in it a symptom of decline.

Einstein declared: ***If we should not sin against common sense, the progress, generally, would be impossible.*** Instead of a conclusion, let me add still one quotation. Luther said: ***Let us sin bravely.***

ACKNOWLEDGEMENT

I wish to thank my non forgotten Barbara for her presence and inspiration during the preparation of this work.

Jagiellonian University
Kraków

NOTES

[1] English is too poor to differentiate the Platonic notion of matter – χορα from the Aristotelian – χυλε.
[2] I think that Poppers 'third word' is kind of over subjective rather, than intersubjective entity.
[3] Shaftesbury, Moralists II, 4.
[4] Emphasis by the author.

BIBLIOGRAPHY

Agassi, J., 'Metaphysics as Regulative Idea of Science', in: *Science et Metaphysique. Colloque de L'Academie Internationale de Philosophie des Science*, Paris 1976.

Einstein, A., 'Remarks to the Essays Appearing in this Collective Volume', in: *Albert Einstein – Philosopher Scientist*, P. Schilpp ed. New York 1959.

Płazowski, J. Werszowiec, 'Czy fizyka jest w pełni racjonalna?' in: *Filozofia dla przyszłości*, Kraków 1987.

Pociej, *B., Bach, muzyka i wielkość*, PWN Kraków 1972.

Polanyi, M., *Personal Knowledge*, Chicago 1951.

Popper, K., 'Epistemologia bez podmiotu poznającego', *Literatura na Ś wiecie* 12/84.

Rayski, J., *Survey of Physical Theories from a Methodological Viewpoint*, PWN Kraków – Warszawa 1978.

MARIAN PRZEŁĘCKI

IN DEFENCE OF THE POSITIVIST VIEW OF SCIENCE

First – a few words about the title of these remarks. They are, in fact, meant not as a proper defence of the accused but as an attempt at understanding the motives of his crime. The accused is the positivist philosopher of science – a being quite rare at the present time (I sometimes wonder if I am not the last of the positivist Mohicans). What I mean by the 'positivist' view on science is also known as the 'empiricist' or simply as the 'received' (or 'standard') view. The view occurs in various forms and versions. In spite of these differences, all the versions have something essential in common. And it is their common content that has become an object of of notorious criticism. There is no need to present the positivist tenets and their criticism on this occasion. Both are well known to anybody concerned with philosophical reflection on science.

In discussing the positivist view of scientific theories, I shall concentrate on what seems to be essential to that view and what determines its further features: on the problem of interpretation of a theory's language. It is just the kind of interpretation of the language of a given theory that distinguishes an empirical theory from a non-empirical (or formal) one. When speaking of the positivist account of empirical interpretation I shall refer not so much to the original conception advanced by Carnap as to some of its later modifications, one of them being presented in my monograph *The Logic of Empirical Theories*.1 The main feature which differentiates it from the Carnapian approach is its semantic rather than syntactic character. Not going into its detailed presentation, I shall restrict myself to mentioning its two fundamental tenets, known usually under the name of the observational-theoretical distinction and the partial interpretation idea. It is those tenets that have become the object of the strongest anti-positivistic criticism. These two principles are closely connected. They amount to the following view of the structure of a scientific theory T.

The specific terms of theory T are divided into two classes: the class of observational and the class of theoretical terms of theory T. It is the observational terms only that are assigned their interpretation in a way independent of theory T. The sole interpretation assigned to the theoretical terms is the interpretation which they acquire through their connections with the observational terms as established by the very theory T. The observational-theoretical dis-

J. Misiek (ed.), The Problem of Rationality in Science and its Philosophy, 195–204.

tinction so understood has a relative character: the observational terms are non-theoretical relatively to a given theory T. In addition to the relative distinction, the classical version of the positivist view assumes an absolute observational-theoretical distinction. According to it, the observational terms are said to be non-theoretical with respect to any scientific theory whatever. Their interpretation is assumed to be assigned to them in a direct, purely ostensive, way.

The criticism of the positivist view so conceived amounts to showing that any account of the structure of scientific theories based on the above principles is grossly inadequate; it gives a false picture of that structure and makes it impossible to explain how a scientific theory functions and how it develops. The division of a theory's terms into observational and theoretical is claimed to be untenable. All the theory's terms are antecedently understood. At the same time, the meaning of all of them is theory-dependent, since their incorporation into the theory alters their previous meaning. It is, especially, the absolute concept of observational terms that is said to be a sheer fiction. No real scientific theory contains terms which are observational in the sense defined, i.e. terms which are interpreted merely ostensively. All these objections to the positivist view are claimed to have been supported by the relevant facts from the domain of science and its history.

In what follows, I do not intend to take sides in the controversy between the positivist philosopher of science and his critics, and to try to decide whether his view of the structure of scientific theories is tenable or not. What I want to do is to look at the way in which positivist philosopher has arrived at his conception and to say why he sticks to it in spite of all the objections raised. The question is understood not as psychological (or sociological) but as a logical one. What are the reasons which speak in favour of the positivist account? What arguments may be adduced in its defense?

In answering these questions attention will be called to certain general assumptions which underlie the positivist approach since it is with regard to those assumptions that the positivist solution is to be assessed. The assumptions in question may be said to define the kind of task which the positivist analysis is meant to realize. Every philosophical conception might be viewed as a realization of a definite theoretical task. Two such conceptions may differ not because they provide different solutions of the same task, but because they attempt to solve two different tasks. This seems to be the case of the positivist and anti-positivist conceptions. Assumptions on which they are based appear to determine different philosophical tasks. Now, what is that task like in the case of the positivist conception? To define a given theoretical task, one has to say, first of all, which questions are to be answered and what sorts of answers are acceptable. As every question presupposes some assumptions, the problem to be solved will be characterized by a question to be answered and by the assumptions which the question turns out to presuppose. In order to explain the notion of an acceptable answer, one should indicate some general criteria (of a more or less formal nature) which such an answer is to satisfy, and determine what kinds of data are to count as evidence for the answer propounded. All these factors involved in the notion of a theoretical task – questions and their assumptions, formal criteria and relevant data – play an essential role in determining that philosophical task which confronts a positivist philosopher. Let us comment on them briefly in order to show how they lead to the positivist solution.

The main factor which determines a theoretical task is the kind of problem to be solved. What kind of problem then is to be solved by the positivist approach? The positivist account of the interpretation of an empirical theory may be thought of as an attempt to answer the question: how is an empirical interpretation possible, or – to put it in a somewhat Kantian wording – how are empirical statements possible. The question so formulated involves the notion of what it is to be empirical, and assumes that a scientific theory is empirical in the sense meant. That sense – rather loose and vague – might be conveyed by saying that the truth-value of a scientific theory is a matter of experience: is is experience that decides its truth or falsity – in particular, it is experience that can show its being false. The positivist account is to explain how a theory can be assigned an interpretation which grants it the empirical status so understood. Since the ultimate aim is to explain the very fact of empirical interpretation, the explanation required must not presuppose that some statements or theories already have an empirical interpretation. It must then be a kind of explanation which might be termed 'radical'(in analogy to Quine's 'radical translation'), as it is to explain the given fact *ex nihilo*, so to say.

To be acceptable, a solution to the given problem has to satisfy some general criteria, which – as suggested above – are one of the factors differentiating various philosophical approaches. They make up what Nicholas Rescher has called 'methodological orientation'.[2] A methodological orientation is said to embody the philosopher's cognitive preferences. It provides certain standards or criteria by which the philosophical theses and arguments are being evaluated. What is then the methodological orientation characteristic of the positivist approach? What kind of criteria does it impose on acceptable solutions? A satisfactory characterization of the positivist 'orientation' is a task for a separate study. Here I shall restrict myself to a few hints only. Most generally, the criteria characteristic of the positivist approach may be said to be those of 'analytic philosophy'. They favor such qualities of philosophizing as its clarity, precision, straightforwardness, its tendency to clear-cut, explicit and literal formulations, to intersubjectively communicable and testable statements and arguments. Those general postulates do not exhaust the standards of philosophizing characteristic of that version of the positivist philosophy which I refer to in the present discussion. Being a continuation of a philosophical trend known as 'logical' empiricism, it aims at what may be called a 'logical' theory of science (in a rather broad sense of the term 'logical', including any formal approach). In the case of the positivist position mentioned above, that tendency manifests itself in adopting the conceptual framework of the standard model-theory. It is in model-theoretic terms that the key notion of interpretation of a scientific theory is couched. The interpretation of a theory's language is identified with a set-theoretic structure called its model, and the question how such a language is being interpreted reduces, on that approach, to the problem of how it is assigned a given model (or rather a given class of models). It is to be noticed that the framework in question is an extensional one; according to it, the interpretation of a given term is conceived not as its meaning (or sense), but simply as its denotation (which seems quite plausible with regard to extensional languages, and such are the languages of most scientific theories). What the positivist philosopher then is looking for is an account of empirical interpretation fulfilling formal requirements of the kind mentioned.

And what is to decide whether such an account is adequate or not? What kind of data are to count as evidence for the solution propounded? The data referred to by philosophers of science may roughly be divided into 'scientific' and 'metascientific' ones. Loosely speaking, the first concern what the scientists are doing, the second what they say they are doing. The first then belong to science itself, the second to what may be called metascience. Besides formulating scientific theories, the scientists, as a rule, make some comments on them. In doing so, they employ such concepts as 'theory', 'observation', 'hypothesis', 'truth', 'meaning', and the like. They act, in this respect, as philosophers of science rather than as pure scientists. The problem arises whether the philosopher of science is bound to respect such enunciations; whether they are to be reckoned among the 'data' of his philosophical theory. The question seems to be answered in different ways by the positivist philosophers and their critics. The latter attach to those 'metascientific' statements much more importance than the former. Taking very seriously the way in which the scientist characterizes his own scientific activity and its products, they aim at an account of the structure of scientific theories which would conform not only to all the 'scientific' data, but also to the 'metascientific' ones. In contradistinction to them, the positivist philosophers are prone to restrict their attention to the 'scientific' data only. They seem to share the belief (expressed by me elsewhere) that a philosopher of science ought to take very seriously what the scientist is doing, less seriously what he thinks he is doing, and least seriously what he says he is doing. His philosophical declarations should never be accepted unquestioningly. Thus, what a scientist considers 'one theory' might well be identified with a part of a theory only or, conversely, with a whole family of theories, and what he calls 'observable' might be reconstructed as highly theoretical, etc. In sum, what the positivist account of the structure of scientific theory (and its interpretation, in particular) is assumed to conform to the body of the 'scientific' data only – first of all, the official exposition of the theory in authoritative text-books or treatises.

The kind of problem to be solved and the kind of solutions sought determine the theoretical task which the positivist is trying to realize. Now, it is my contention that if one is to realize the task so conceived, he is bound to arrive at something like the positivist view of the interpretation of scientific theory. The contention can hardly be justified here. What I can do now is to restrict myself to a few suggestions and to refer to other works for a more detailed argumentation.[3]

The assumptions characteristic of the positivist approach find expression in a principle known as the Thesis of Semantical Empiricism, which seems to be at the bottom of the positivist conception of an empirical interpretation. According to that thesis, there are only two ways of providing an empirical interpretation for a given term: a direct and an indirect one. We may assign to a term its denotation directly, by pointing it out, or indirectly, by describing it by means of some sentences (called postulates) which it is assumed to satisfy. These are claimed to be the only interpretive procedures which can be identified in a sufficiently clear, explicit and straightforward way. Now, it is easy to show, that if the indirect procedure is to supply a term with an empirical interpretation, the postulates must contain, besides the term in question, some terms already endowed with empirical interpretation. And so, to furnish a term with an empirical interpretation, one has ultimately to resort to a direct procedure. These ascertainments can be shown to entail that much discussed feature of the positivist conception which is known as the observational-theoretical distinction, taken in its relative sense.

If to be dependent on a theory T is to satisfy the axioms of theory T, not all terms of theory T may be claimed to be T-dependent. If that were the case, the theory T would not be empirically falsifiable. Its interpretation would be defined in a way which guarantees the truth of theory T in advance – independently of any empirical findings. In effect, the theory T could never be false: it would be either uninterpreted or true. That is why the terms of a theory T are being divided into dependent and independent of theory T – that is to say, into theoretical and non-theoretical (or observational) with regard to theory T. This is a relative distinction only. The terms non-theoretical with regard to theory T may well be theoretical with regard to another theory T′, which underlies theory T. A philosopher who, like the positivist, aims at a 'radical' explanation of the fact of empirical interpretation cannot let the matter rest here. To insist that any scientific theory presupposes another scientific theory recalls a well-known argument about machine tools. To make a machine tool, another machine tool is needed; *ergo*, the world must have been created already with some machine tools. In order to avoid a vicious circle or an infinite regress, one has ultimately to resort to something like 'basic' empirical theories. The non-theoretical vocabulary of such theories is assumed to consist of terms which may be called observational in the absolute sense, their interpretation being independent of any scientific theory, not only of the given one.

I cannot go into any details of the positivist conception on the present occasion. To guarantee a scientific theory an empirical status, its interpretation must satisfy some further conditions. If theory T is to be empirically falsifiable, its theoretical terms must be dependent not on the whole set of its axioms, but only on a certain part of it, identified with the definitional (or analytic) component of theory T (its role might be played by the so-called Carnap-sentence, though this is not the only solution). That kind of interpretation of the language of a scientific theory ensures the theory an empirical status understood in a rather strict sense. It subjects scientific theory to the control of an independent tribunal of experience and, in effect, enables rational choice between competing theories. Two such theories may well be comparable and logically incompatible due to the fact that they are assumed to contain observational vocabulary interpreted in a way independent of either theory. And so, the choice between them may be determined by the outcomes of 'crucial experiments'.

It is that kind of view of the interpretation of a scientific theory that has become the object of the anti-positivist criticism mentioned above. Criticizing it, its opponents have propounded some alternative solutions, which are claimed to be free from the positivist shortcomings. Now, how do the defenders of the positivist view cope with the objections raised? And how do they assess the alternative proposals? As said before, the positivist view is claimed to be consistent with the 'scientific' kind of evidence concerning the given theory. And what the critics of that view may be said to have shown is its inconsistency with the 'metascientific' kind of evidence – with certain philosophical comments on the given theory as made by the scientists themselves. This is claimed to be true of such fundamental features of the positivist view as the observational-theoretical distinction and the partial interpretation idea. It is contended that a mere exposition of a scientific theory does not exclude its interpretation based upon those tenets in view of the fact that the scientific context leaves the notion of a scientific theory fairly indefinite.

One of the objections directed against the observational-theoretical distinction taken in its relative version refers to the fact that the same term of a given theory may function in some contexts as a theoretical, in others as a non-theoretical one. The usual way of coping with that objection amounts to claiming that in a situation like this we have, in fact, to do with two different (though closely related) scientific theories and not with one such theory – contrary to the scientist's statements. But, as mentioned before, it is the absolute version of the observational-theoretical distinction that has been most strongly criticized: no scientific theory is said to contain observational terms so understood.

Answering this objection, the positivist philosopher maintains that whether the above statement is true or false depends on what kind of entity a scientific theory is supposed to be. On certain narrow conceptions of scientific theory most such theories do not contain observational terms strictly understood. Let us take classical particle mechanics as a typical example. According to its current conception, it is a theory comprising Newton's laws as its only physical axioms and the function symbols: 'position', 'mass' and 'force', as its only physical terms. None of its terms, of course, is an observational term in the absolute sense. The same theory, however, may be conceived in a more comprehensive way: as including all the subtheories which underlie it – in particular, all the theories of measurement of its fundamental physical quantities. Since some of them may be identified as 'basic' theories in the sense mentioned before, the theory so conceived will contain something like observational terms strictly understood. When we carry an analysis of measurement procedures far enough, we are bound to reach some terms (say, 'coincides') which might roughly be identified with observational terms in the absolute sense. A strict identification is, perhaps, never possible. If it is assumed by a philosopher, it is assumed by him as a kind of idealization.

The positivist philosopher considers his comprehensive view of scientific theories as an admissible one. The scientific context does not determine uniquely what exactly belongs to the given theory and what does not. It has to be admitted that the positivist conception of a scientific theory does not conform to the scientist's linguistic customs. But a philosopher is not obliged to obey his way of speaking. It has also to be admitted that including into a given theory all theories of measurement underlying it, the positivist philosopher replaces a well defined, neat structure by a vague and unwieldy whole. This approach, called by John Watkins 'unmanageable largism' and considered by him to be destructive for a theory of corroboration (or rather for its application),[4] does not seem to be destructive for a general conception of empirical interpretation, which does not involve an actual reconstruction of concrete scientific theories. Consequently, the positivist philosopher does nor regard his view of the interpretation of a scientific theory as inadequate. Being based on an acceptable (though unusual) conception of of a scientific theory, it is claimed to provide an admittedly idealized, but approximately adequate account of its interpretation.

The reaction of the positivist philosopher to the alternative solutions proposed by his anti-positivist critics is determined by the kind of task which he is trying to realize. It is, first of all, the kind of problem to be solved that makes some of those proposals unacceptable to him. In order to avoid unrealistic assumptions in his view of a theory's interpretation, the anti-positivist philosopher does not hesitate to refer to an 'antecedently understood' scientific vocabulary – without any explicit account of how that vocabulary has come to be understood. That kind of assumption is, of course, unacceptable to the positivist philosopher, who aims at a 'radical' explanation of the very fact of empirical interpretation. Another version of the same anti-positivist strategy – of a more formal nature – has been represented in some recent works on the subject. The author of one of them is prone to identify a scientific theory as a partially interpreted calculus, rejecting at the same time the positivist analysis of that interpretation. In view of the extreme complexity of the interpretative procedures, he proposes "to represent the process of interpretation abstractly as an unspecified function from sentences to truth values".[5] This, again, means abandoning an attempt at what I call 'radical' explanation.

There are, of course, anti-positivist philosophers who do not assume some kinds of empirical interpretation as something given, but try to provide an exhaustive analysis of a theory's interpretation. Their solutions, however, are hardly acceptable to the positivist philosopher, since they do not satisfy his formal criteria of acceptability. The notions and procedures which anti-positivist philosophers resort to are, as a rule, much more involved and enigmatic than those characteristic of the positivist approach. When speaking of interpretation, they make use of the concepts of meaning (or sense) rather than denotation. When describing the ways in which the meaning of a term may be determined, they appeal to procedures such as "recourse to analogies and iconic models", and the like. Assuming the theory-dependence (or theory-ladenness, as they usually call it) of all theory's terms, they do not define it in the same straightforward way as on the positivist approach; instead of logical, they characterize it rather by psychological notions. Such an account can hardly be submitted to a precise formulation, let alone formalization. In effect, most of the anti-positivist conceptions are expressed in a suggestive, perhaps, but fairly loose manner, being often couched in metaphorical rather than literal language.

These differences reflect differences in the general approach to the issues discussed. Any anti-positivist philosopher seems to conceive his main task in a way which differs from that pursued by his positivist colleague. He does not aim at a 'radical' explanation of a theory's empirical status. What he, first of all, aims at is such an account of a theory's structure which is fully realistic: applying to real scientific theories and agreeing with actual scientific practice. And, as I have put it before, it is to be an account compatible with the 'scientific' as well as the 'metascientific' data. What is more, his analysis of theories is to include the developmental aspects of scientific theorizing and not to limit itself to a logical reconstruction of theories at fixed stages of development. The task so conceived is exceedingly difficult and can hardly be accomplished in a way precise and adequate at the same time. In this situation one faces a choice between comparatively precise but highly simplified construction, or more or less adequate but rather vague generalities. In contrast to the positivist philosopher, his anti-positivist critic appears to prefer the second alternative, in accordance with his intellectual predilections.

Some statements of the anti-positivist philosophers suggest that they are presenting a different solution to the same problem which the positivist philosopher has been trying to solve, and they manage to do it without accepting any controversial positivist assumptions. It is my contention that this is not the case. The positivist philosopher gives an answer to a question which his anti-positivist critics leave unanswered. And so, whatever shortcomings his solution might have, it should not be rejected too easily. With respect to a genuine theoretical question, a wrong answer is better than none, because it presents a challenge to further inquiries and improvements.

Institute of Philosophy
University of Warszawa

NOTES

[1] M. Przełęcki, *The Logic of Empirical Theories* (Routledge and Kegan Paul, London 1969).
[2] N. Rescher, 'Philosophical Disagreement: An Essay Towards Orientational Pluralism in Metaphilosophy', *Review of Metaphysics 32* (1978), 217–251.
[3] See note 1.
[4] J. Watkins, *Science and Scepticism* (Princeton University Press, Princeton 1984)
[5] B. Mundy, 'Scientific Theory as Partially Interpreted Calculus', *Erkenntnis 27*, (1987) 173–196.

WILLIAM T. SCOTT

ON POLANYI'S NOTION OF RATIONALITY

Michael Polanyi was born in 1891 in Budapest, the fifth child of a brilliant family of the educated Hungarian elite, where the two traditions of 1848 Liberalism and pre-war Marxism and the new thought from Western Europe all competed for the allegiance of the intellectuals. Polanyi started his career with a degree in medicine and joined the Austro-Hungarian army in September 1914. Taking care behind the front of the victims of infectious diseases, he succumbed himself to diphtheria and while recuperating back in Budapest wrote an extensive and innovative paper on the adsorption of gases on solids. He first published it and then converted it into a Ph.D. thesis in physical chemistry, finally getting his degree at the war's end.

In due course Polanyi made his way to the Kaiser Wilhelm Institute in Berlin where he achieved fame for work in X-rays and crystals, in the rates of chemical reactions, and continued work in adsorption. In 1933 he accepted a post in chemistry at Manchester, England and with his wife, his son George at 11 and his son John at 4 pulled out of the town and country they had learned to love and started life over in a new place.

By this time his deep concern about the Soviet Union had begun to take shape and started to interfere with his work in chemistry. Why could the Bolshevik revolution be so inhuman, so in pursuit of an impossible ideal of government, so anxious to dispose of the values they called 'bourgeois morality,' and still have the audacity to claim to base their rule on science? He published his first book on Soviet government, *USSR Economics*, in 1935. In study of capitalist economics, he became convinced that the evils of capitalism that were said by socialists to spell its doom were curable. What was needed was education of the public to make clear the principles to follow in the reform of capitalism. He was especially concerned with the problem of unemployment accompanying depression. He made many efforts to educate the public by numerous lectures and a film he created on the circulation of money. At the same time he wrote extensively on the foundations of a free society, pointing out the inefficiency, backwardness, and cruelty caused in the USSR by not building on these foundations. In 1948 he moved from the Manchester University chemistry department to the department of social studies and started his philosophic search in answer to his basic question: Where have we in the

J. Misiek (ed.), The Problem of Rationality in Science and its Philosophy, 205–214.

West gone wrong in our views of science that we would encourage both in the Bolsheviks and in ourselves the dehumanizing aspects of the prevailing ideologies and social passions? His search led him to uncover the basic features of knowing that he laid out in his great book *Personal Knowledge* and other works, forming a new vision of rationality in nature and a new view of how we search for and come upon such a vision. Polanyi retired from Manchester University in 1958 after completing his book and moved to Oxford, from which he traveled to many speaking and teaching engagements. He died in 1976.

My quest for a way to put together my love of physics and my social concern for a world gone mad with its inhumane scientism found its true road when I met Michael Polanyi at a lecture in 1959 and read his then just published *Personal Knowledge*. It is my intention in this article to give in a Polanyian framework an account of several elements of the experience of scientists participating in the search for rationality in the world, hoping it will be as persuasive to you as Michael Polanyi intended it to be to his readers. I shall start with reference to Schrödinger about whose philosophy I have written.

Erwin Schrödinger described the origin of science in Pre-Socratic Ionian Greece as occurring with the acceptance of the two principles of comprehensibility and objectivation, which I take to be basic to Polanyi's, Popper's or anyone's notion of rationality. Comprehensibility is the assumption that all the happenings in the world can be understood in rational terms. It is the non-superstitious, the non-animistic, the non-magical outlook. The principle of objectivation says that to find the understandable reasons for anything we must at least temporarily stand apart from it and treat it as something real out there. Polanyi is a realist and it is important at the outset to say what he means by something being real: it is that which may yet inexhaustibly manifest itself, that which we expect to meet in unexpected ways.

The Greeks searched for the reasons for happenings using a wide variety of modes of thought, both the analytic or deductive type arguing from general principles to specific phenomena and the integrative or holistic type, finding explanations in terms of structure, function and purpose. Modern science is still a matter of thinking of the world in the Greek way, but its philosophy has mainly come out of the Enlightenment which narrowed its view of the Universe and the mode of its explanation almost exclusively to deductive arguments inspired by the certainties of mathematics. Holistic thinking was generally associated with unprovable and often unaccepted matters of religious faith. Descartes is credited as the originator of the effort to purge philosophy and science of unfounded beliefs by subjecting all assumptions to a process of radical doubt, keeping only the completely indubitable. The method was not successful in spite of Kant's supporting efforts over a century later. However, his so-called Critical Philosophy remained as part of the Enlightenment, following the ideal of science modeled on the apparently universal certainty of the Newtonian mechanical philosophy.

The rejection of holistic thinking meant that most of human life and culture was omitted from this mechanical outlook. In response, a variety of writers in what is loosely called the Romantic Rebellion tried to restore to respectability the ethical, spiritual and artistic sides of life. The authority of the mechanical view was too great for these romantics to attempt to take it into their thought – an outright break with it was essentially all they could do.

The importance of Michael Polanyi in this picture is his demonstration that holistic thinking is fundamental to all science including the theory of Newton and its successors, so that far from making a cut-and-paste job of two apparent contraries, he reveals the basic integration of science and human life.

To give a preview of Polanyi's transformation of the mechanical philosophy, let me suggest four basic consequences of that philosophy and what Polanyi did to get beyond them. The successful application of Newton's axioms and his law of gravitation to the solar system was extrapolated, notably by Laplace, to a systematic Democritean view of all things being made of atoms exerting law-obeying forces on each other. The first consequence of this system is deductivism, the view that any area of science can be described in terms of a few indubitable axioms from which any of the possible consequences can be derived by mathematical logic. The second is determinism, coming from the specific character of the mechanical deductions that gave precise prediction of future motions. All happenings of any sort in the world were asserted to be in principle predictable and therefore determined. The resulting ideal of science as aiming at axiomatic systems capable of prediction was exchanged by Polanyi, among others, for the aim of understanding nature, that is of searching for rationality in the real world, which means in general terms uncovering the entities and mechanisms by which things happen, and the connectedness of things with each other, none of which can be derived from the Laplacean outlook. While Polanyi considered the theory of the solar system as a grand example of rationality, a great deal of the science and other knowledge he treated is not axiomatizable, and the first theorem was not of great importance. Secondly, Polanyi's view of personal action has creativity and novelty built into it everywhere, so that determinism also has little place in his view.

The third consequence of the mechanical philosophy is positivism, the idea that the only observations that provide any true knowledge are those that can be made precisely and impersonally and have a bearing on the state of motion of an object or better of its atoms. We shall see that impersonal measurements do not exist for Polanyi and that varying degrees of holistic thinking and personal judgment enter into all of them.

The fourth consequence of the Newtonian view is the claim of reductionism. The atomistic philosophers thought that it was possible to reduce all the laws of complex things to the laws of physics, and their philosophy depended crucially on this idea. Polanyi rejected reductionism both epistemologically and ontologically by a view of the world as made of many independently real levels or strata possessing among them a wide variety of rationalities.

Polanyi's philosophy is centered on holistic thinking, in particular on an elementary conception of how such thinking occurs, how the thinker is involved, and the ubiquity of the process. The basic power of understanding involves obtaining a vision of the joint sense made of many aspects of a situation. It stems from a physiological urge inherent in our sense organs which impels them to comprehend rationally the impressions they receive. This power we know in exercising it but find difficult to talk about because we attend to the content, not the process. Polanyi calls it the power of tacit knowing that we obviously have and equally obviously cannot describe explicitly.

The simple example of perception of an object can be used to illustrate Polanyi's concept of tacit knowing. We start with the fact that we receive many bits of light and shadow from the object, and have many pieces of knowledge in our memory that help us to form our perception. Nevertheless, we do not perceive a world of bits of light and shadow, nor fluctuating pieces of information from our memory. What we see are things with definite shapes, places, motions and meanings – Gestalts, to use a general term.

What is novel in Polanyi's description of the process are two points that Polanyi took from the Gestalt psychologists. First, when we see the Gestalt we rely on all the bits and pieces but do not look at them. We see them in a fundamentally different way than the Gestalt on which we focus our attention, that is we see them subsidiarily as pointers to the Gestalt. If we tried to look directly at the bits we would lose the Gestalt and they would stop being pointers to it. We don't know consciously what most of the pointers are, and if for instance we would guess what features tell us that the Gestalt is our friend Paul, we might guess what we used yesterday as pointers but not those particular ones we are using now. Clues such as these cannot be explicitly talked about; insofar as they must be part of our knowing of Paul, Polanyi says they are 'tacitly known.' He also used tacit knowing for the combination of focal and subsidiary awareness that participate jointly in perception. The second aspect of the tacit is that no one can tell how the clues go together. Tacit integration is Polanyi's name for this process, for which there is no good theory although psychologists can measure its speed, happening in less than a second in most cases.

The same process of clues and integration, of things we know and cannot tell, underlies the recognition of meanings in language, the grasping of new concepts, the comprehending of visions of rationality in nature, and all of our mental life. We use our language in an enormous variety of ways as we develop and live in the culture which separates us so far from our pre- lingual ancestors, and it is our tacit powers that guide this use at every step. For instance, these powers guide the critical reflection in explicit writing that is responsible for the enormous store of accepted cultural and scientific wisdom in the libraries of the world. The fact that our tacit powers are not themselves subject to critical reflection is an expression of the fundamentally personal underpinning of language and culture.

Tacit knowing is a skill that we learn by practice. It has a close relation to bodily skills which involve focal awareness of what we are trying to do and subsidiary awareness of all the components of our action. Since we cannot critically reflect on any of these components of knowing and doing, we necessarily have to trust ourselves and equally to trust the skills of others with whom we interact. Knowledge is personal in the first place in respect to the existence of the tacit component and in the second place in respect to the necessity of trusting this component.

Let us look further at the two kinds of awareness as they function with bodily skills. Consider the use of a tool such as a hammer or a pen. The objects of our focal attention, nails to be driven or words to be put on the page, are out there where we can relate to them. But the tools are attended to only in the subsidiary mode of reliance for the action in question. The sensations of fingers on the tool are not noticed as such but are utilized for guidance of the tool. The tool acts as a part of our body. I could have used my hand instead of a tool for the example, but the two points I wish to make are more obvious for the tool. The first point is that our body with or without extensions is known almost exclusively in the subsidiary mode – we rely on it for getting about in the world. Secondly, all the particulars that we integrate in any tacit knowing are in ourselves – they are the sensations, such as those on the retina which had formerly been thought of as forming images that we somehow could observe focally, and all the memories and beliefs we include with the subsidiaries. In accordance with the common-sense expression to the effect that we dwell in our bodies, Polanyi says that we dwell in a group of particulars pointing to a given Gestalt, including the tools we assimilate to our body. Thus a third way in which knowledge is personal is that all our tacit knowing is embodied.

This embodiment represents the subjective aspect of our relation to a Gestalt. When we take in the particulars and become convinced that what we see or conceive has significance as being true or plausible we commit ourselves to that perception. That is, we move toward it, we take a position to hold to the conviction, to rely on it and act on it. It does not happen without motivation, and the term commitment implies a passionate motivation – the whole area of values in science and in life are discussed by Polanyi under the rubric of Intellectual Passions.

Commitment has a polar nature in that the inward end of the pole is grounded in our subjective indwelling and the outward end is grounded in the reality to the existence of which we are committed. If we were to suspect that our subjective preferences biased our thought, they would threaten the universal intent with which we committed ourselves to something or other as being real and we would change our commitments. Polanyi uses the term 'personal' to bridge the gap between subject and object. He does not reduce the power of object as object, but adds to the power of subject by its participation in the personal.

Another important factor of tacit knowing is that it is transitive, in that we often shift attention from a single Gestalt to a group of them and find a new Gestalt which may be an entity or a meaning composed of the members of the group and of aspects of their situation which now become particulars. We go from one level of reality to a more complex or more meaningful one. The world meets us as a vast set of differing but interlocking levels of reality and rationality, all calling out our commitment as we get to know them. This is the epistemological ground for Polanyi's conception of stratification.

Many of our visions of rationality are of organized structures, for which we postulate or perceive principles of operation, of function, of purpose, or of order. The status of these principles is partly a matter of our epistemological choice of what level to look at, but more fundamentally of ontological irreducibility to parts obeying laws at lower levels than those of the observable organizing principles.

Whether the parts under consideration are atoms, or levers and gears, or computer chips, or animal muscles, the behavior of each is described by laws that can be formulated with sufficient generality that they cover any circumstance in which the part can function without losing its individuality. On the other hand, accounting for the behavior of a given collection of parts in given circumstances needs information not only on these laws but on the specific application of them, that is, on initial and boundary conditions, specification of numbers and arrangements, and other elements all of which Polanyi calls marginal control. It is marginal control that determines that a set of gears and levers or of computer chips and liquid crystals form a watch or a calculator or a control device or something that doesn't do anything at all. The central point is that the rational organizing principles of the object are particular laws of marginal control which are logically excluded from the general laws of the parts. The parts of something are often composed of smaller parts, so their general laws are marginal principles for the smaller elements. Conversely, the entity in question may have a function as a part of a larger, logically more complex entity.

I find it convenient to refer to an ordering, organizing or operating principle of a given entity by the letter *O*; and the underlying laws of any of its parts by letter *U*. We then can describe the relation between the whole and the parts by saying that the *U*'s are necessary but not sufficient for the *O*'s. Rationality for a machine or for machine-like animal functions does not become evident by a theoretical process of reduction, which would amount to setting the marginal control principles aside as not of explanatory power, but rather by spelling out the marginal control laws in terms of the design purpose of a human invention or the operational function of a feature of a living organism.

In Polanyi's stratified view, the Universe consists of an enormous interlocking hierarchy of organized things, happenings, and ideas, including most preciously the vast variety of products of culture. At the bottom of the hierarchy is the realm of the particles of physics. The entities concerned here are few in kind and exist in vast numbers so that widely applicable generalities of axiomatic form are the natural subject matter. But as we go up the scale of complexity, the elements become more and more variegated with fewer and fewer individuals of any given kind, so that the rational sort of generality in physics becomes less and less useful and the specificities of organized complexity become the vehicle for understanding.

The wide variety of entities and structures in the world finds its reflection in the wide variety of sciences, and the interlocking relations of the entities correspond to the interlocking nature of the sciences. A point to notice in the hierarchy of *U*'s to *O*'s which then become *U*'s for the next set of *O*'s and so forth is that an *O* may entail a higher or a lower degree of determinism than its *U*'s. The higher levels of being human entail more freedom and creativity than is provided by the determiners of the living machines of our bodies. On the other hand, a computer has far more determinism than can be seen at the Heisenberg level of the wave-like electrons which computer chips utilize.

Since no scientist can find by experiment more than a miniscule fraction of the information he or she needs for research, it is vital that she or he be a member of a research program and that the scientific community provide the communication needed for such organized groups and even more importantly form a system of publication and other communication to provide information that can be trusted – still another personal aspect of knowledge. The interlocking nature of the sciences leads to an enormous web of critically judged scientific opinion which maintains the continually modifying standards and premisses of all the disciplines. Rarely is a published result tested by repetition, since the criteria by which a researcher and the appropriate referees consider something likely to be a genuine discovery are the same as the criteria for verification. It is a fact that accepted publications generally bear signs of adequate verification. Each result fills in some sense a gap in the growing web of science. One of the important types of rationality for Polanyi is the rationality of newly discovered coherences with that web, a part of the general connectedness of things.

Innovation and discovery normally go on in science through the choice of good problems, which we could characterize as the gaps for research that appear at the edge of this continually growing web of scientific thought. Tacit knowing used in judging a problem to be a good problem involves an integration of known results in the nearby part of the web with available resources and the particular skills of the researcher, all into a focus on the unknown, like trying to look through a thick fog. Once started, the research begins to clear the fog, more clues appear and the new result emerges, quite rational but often looking rather different than the first vague guess. This process of research and discovery is the general character of the way rationality is found in the scientific world. Research programs are shared visions by small groups of scientists of aspects of reality that inspire a steady succession of good problems.

The conceptually difficult part of Polanyi's theory of knowledge and the most deeply personal is what he calls his 'Fiduciary Programme,' the problem of the acceptance of unproven beliefs that the critical philosophy was supposed to eliminate. If we consider the arguments that an individual may use in deciding what to believe of his own results and the arguments that the scientific authorities use in propounding the state of the art in their respective fields we see that they rest on a number of beliefs that we all imbibe from our culture but as the basis of our provable beliefs are not themselves provable. They are matters we could conceivably doubt but do not actually doubt; they are more elements we take on trust; we commit ourselves to them.

For instance, we believe that we live in a real world with which we can make contact. We believe that our mental powers are adequate for making sense of our experience and that it is valid to use the terms of our own adopted standards of rationality. We believe that scientists by and large are competent to follow their impulses in pursuit of discovery. We believe that our communally shared passions for beauty, elegance, coherence and truth are trustworthy guides to these pursuits.

We believe that the interpretative framework embedded in our language is largely a true one, and that a large majority of the results of scientific research are valid. We believe that we can recognize correct logic, geometric implications, the existence of order, and the existence of statistical randomness in the physical world. I should add my own conviction that scientists can recognize coincidences that are genuine accidents and not subject to rational explanation.

If all assertions of truth rely on unproven beliefs, how can there be knowledge? Do we have to be satisfied with mere belief, dependent on the individual's particular preferences, or can there be genuine contacts with reality that are asserted and accepted as true? In the first place, truth can only be talked about by believing it; otherwise we have only alleged truth to refer to. Many of the truths of our civilization are shared so widely that their assertion takes on impersonal form and they may appear to stand by themselves without being asserted by someone. Nevertheless personal assertion is all we have. Since anyone bears the risk of being in error, the form of acceptance can only be a commitment. In a commitment the person relies on all the available clues suggesting that what is perceived at the universal pole is indeed a contact with reality. Taking the hazard of possible error into oneself, one concludes it is time to stop questioning or doubting, accept the piece of knowledge and dwell within it, adding to one's personal store of knowledge to which one trusts one's mental existence.

Once recognized, the existence of the set of unproven beliefs behind the commitment appears to threaten the person's convictions of truth. The specter of cultural relativism raises its head. Would a person coming at this material from a different culture find a different truth? If in the desire to remove the doubts thus raised one tries to conceive of finding a standpoint from which to compare, say, our Western framework of belief with some traditional Asian or African one, one will attempt in that comparison to make judgments on such matters as the question of rationality. Immediately one is back in one's Western tradition which has a well-defined concept of rationality and standards for recognizing its denial. One cannot actually find a place to stand that might let in a different frame of reference because the search is governed tacitly by the very beliefs that draw one to ask for a more reliable basis for one's conceptions of truth.

In short, there is no one here but we fallible human beings doing the best we can to explore reality, and discover rationality, and we are convinced that in spite of the risks we have done pretty well.

To summarize, Michael Polanyi's notion of rationality is based on integrative thought as the means whereby persons recognize rationality in nature. The concept of the irreducibly stratified universe implies that there are many kinds of rationality known in the various sciences and countless more are awaiting discovery. A non-exhaustive list of particular types includes: the purely deductive systems in mathematics, the deductive systems concerned with elementary entities in physics, geometric systems such as crystal physics and organic chemistry, structural-functional systems in biology with many kinds of organizing, ordering and operational principles, engineered systems combining *O*'s and *U*'s, many complex systems of thought, relation and action in the social sciences, and quite generally, the filling of gaps in the evergrowing edge of the web of science.

These examples of rationality in nature are the sort that satisfy the intellectual passions of scientists and scholars and strengthen their faith that there is an inexhaustible supply of more cases to be found. Michael Polanyi's conception of the personal as bridging the gap between the subjective and the objective is exemplified in each case.

Department of Physics
University of Nevada
Reno

MICHAŁ TEMPCZYK

LIMITATIONS OF DETERMINISM IN CLASSICAL MECHANICS

The search for a sure explanation and prediction of phenomena is the main aim of science. Such an explanation can be obtained on the basis of deterministic laws and deductive reasoning only. On the other hand it is well known that in many fields of science deterministic description is impossible, so one has to use statistical laws which instead of exact predictions give probabilities of individual results. Therefore the application of statistical laws in science has stimulated many discussions, and numerical attempts at epistemological justification of them have been success less.

Before the rise of quantum mechanics it was widely believed that fundamental physical processes obeyed deterministic laws and physics was an ideal of scientific explanation. However, quantum theory is irreducibly statistical and that fact is still a serious problem for the philosophy of science. The sixth chapter of of *Science and Skepticism* by J. W. N. Watkins begins with a paragraph entitled "The Challenge of Microphysics". The author writes:

> I hold that here is only *one* serious challenge to the deductivist ideal for scientific explanation, but that it is *very* serious. Many of the theories in contemporary physics that conform admirably with the of (B*) for depth and predictive power have an essentially *statistical* or probabilistic character; and it is widely held that such theories, although they explain much, cannot provide *deductive* explanation for what *they explain.*[1]

Watkins assumes, in accordance with prevailing opinion that classical mechanics is determinist.

My aim is to show that, according to recent results of the abstract theory of dynamical systems, the role of determinism in classical mechanics is smaller than philosophers believe. The problem of statistical and deterministic laws and their interrelation appeared in physics in the second half of the nineteenth century, in connection with the rise and development of statistical mechanics. In that theory, fundamental interactions of a system of elements are mechanical, so they are determined and reversible, whereas the whole system develops in statistical and irreversible way. The problem was how to explain the origin and the nature of irreversibility and statistics in such a system. Many physicists and mathematicians have worked on that problem but during a hundred

J. Misiek (ed.), The Problem of Rationality in Science and its Philosophy, 215–219.

years they have failed to achieve its complete explanation. At the beginning it was believed that the deterministic theory (classical mechanics) described the nature of phenomena, while the statistical description was the result of scientists' ignorance. The statistics appeared when some degrees of freedom of the systems were neglected, when physicists passed from detailed microscopic to macroscopic description, when the procedure of coarse graining was used, etc. Summing up we can say, that in the course of one century determinism was treated as essential and statistics as superficial characterization of physical phenomena. Many paradoxes which appeared in those times were not fully solved.

The situation changed in the last thirty years, although we owe the first interesting results to Poincaré, who initiated the study of the mathematical stability of classical dynamical systems. He observed that the well-known school examples of mechanical processes, such as planetary motion, collision of balls, the harmonic oscillator etc., describe motions which are exceptionally stable. It means that if in the starting point two trajectories differ slightly, their difference remains small in the future. In other words, small errors in the measurement of the system's position and velocity lead to small changes in its future trajectory. That fact is of fundamental importance in practical applications of science (e.g. astronomy). It is called stability of motion. The stability depends on the kind of differential equation and a potential, both describing the class of processes. In classical mechanics the equation is universal, so the stability depends on the potential only.

Poincaré examined the problem in a general way. He defined an abstract space of dynamical systems and a measure on that space. He also proved that stable systems form in the general space a 0-measure set, i.e. almost all systems are unstable. It was a very important result giving physics new perspectives. The theory of dynamical systems was continued by Birkhof, Hopf and many others, who considered the first degree of instability – ergodicity. Ergodicity means that time averages and phase averages are equal in a system. Detailed studies of ergodicity lasted for decades. The main problem was the classification of ergodic systems. In 1958 Kolmogorov defined entropy as an invariant of automorphisms.

Kolmogorov's paper was the starting point of a new exciting period of rapid development of the theory. Mathematicians have defined several levels of system instability: ergodicity, mixing, K-flows, Anosov systems. Ornstein has proved that two Bernouli shifts with equal entropy are isomorphic. The mathematical theory of unstable dynamical systems uses many advanced methods and concepts, such as the spectral theory of operators, the measure theory, entropy etc.

For philosophical discussions concerning the differences and connections between deductive and inductive explanation, the crucial of this mathematical theory is a new picture of deterministic and statistical systems. It has been proved that many deterministic systems, in many aspects behave like statistical ones. If a deterministic system is unstable, it is similar to a statistical system. For unstable systems with nonzero entropy, similarity means mathematical equivalence. In other words, the determination of the statisticity of unstable systems depend on the point of view and the taste of the researcher. There is no essential opposition of determinism and stochasticity if the systems considered are unstable. The instability of motion causes deterministic systems to look like statistical ones. Mathematicians have proved that well-known dynamical systems – geodesic motion in a connected space with negative curvature, billiards, three-body systems, the Lorentz gas – are such systems. They are classical deterministic systems but there is no possibility to separate individual trajectories.

Geodesic motion in a negative curvature space is the most striking example of such a case. It is an inertial motion, there is no interaction but the geometry of the space results in an exponential divergence of trajectories. Small differences in the starting point give essential changes in trajectories. Similar situation appears in billiards. If two balls starting from one point with equal velocity differ slightly in directions, than their their trajectories may be completely different.

The transformation from a deterministic to a stochastic description of motion is reversible. There is no loss of information nor approximation. Both descriptions contain the same information. It is the reason why scientists speak about random or stochastic motion. Because of the instability of motion, the system soon forgets its initial state, so it behaves like a memoryless system. The consequence of instability is that there exists a finite probability of coming from any small region of the phase space to any other region and, so we face here a typical statistical behavior.[2]

What are theoretical and philosophical consequences of that fact? First of all, the deterministic-statistical opposition, fundamental for science and philosophy, disappears. In many kinds of motion, instability excellently imitates stochasticity. If the instability is of sufficiently high degree (nonzero entropy) the deterministic and statistical descriptions cannot be distinguished, they are mathematically equivalent. A quotation from a physicists' paper describes that situation:

> The common feature of dynamical systems having a suitable high degree of instability is

> that each finite region of phase space, no matter how small, contains phase points that move along rapidly diverging or qualitatively distinct types of trajectories. Obviously, in this situation, the concept of deterministic evolution along phase space trajectories cannot be defined operationally and hence, constitutes a physically unrealizable idealization. Therefore, in dealing with dynamically unstable systems, classical mechanics seems to have reached the limit of applicability of some of its own concepts. This limitation on the applicability of classical concept of phase space of trajectories is – it seems to us – of a fundamental character.[3]

The above conclusions concern the conceptual scheme of physics, but they have wider relevance. The same physicists write farther:

> It seems to us that the significance of this result extends far beyond its immediate application in statistical physics. It proves, of course, Boltzman's H-theorem for the Bernouli systems without invoking the questionable 'molecular chaos' hypothesis or any form of coarse graining. But more importantly, it confirms our view of how probabilities could arise in physics other than a result of approximation or inexact observation. . . In fact, the very possibility of such a connection between deterministic and probabilistic description (which is established in the specific case of Bernouli systems) is rather unexpected result which puts the entire problem of the role of probability in physics in a new perspective.

This quotation emphasizes the role of probability in physics, but it can easily be generalized to the role of probability in all science and the philosophical concept of probability. Until now there seems to have been an essential difference between deterministic and statistical systems and accordingly one can decide which kind of system he deals with. It needed only to make sufficiently precise measurements or observations. However, the mathematical equivalence of unstable deterministic systems and statistical Markov chains shows that in such cases the inclusion of a system into one of these categories is a matter of convention and free decision. In practice, each observed system is known with limited precision, so we never know individual trajectories. We cannot decide finally whether the phenomena examined are really determined or statistical. Brownian motion is a good example of such a situation. On the one hand it is a classical model of diffusion i.e. a continuous Markov process, but on the other hand it is known that it is a result of deterministic mechanical collision of molecules. It was earlier believed that more and more precise measurements of those collisions would bring physicists nearer to the ideal of unique and deterministic knowledge. Now we know that such knowledge is impossible to reach.

The argument of the mathematical equivalence of instability and stochasticity of motion can be used in both directions. On the one hand one can claim that even if the system is essentially determined, as in the case of Lorentz gas or billiards, the initial information is quickly dissipated and the whole system approaches the most probable state. In practical scientific activity, almost all systems are therefore statistical and the level of statistical description can not be overcome. It is neither the result of our ignorance nor inaccuracy of measurement. The fact is a serious challenge to the deductivist ideal for science, even more serious than quantum mechanics, since classical mechanics has been considered as an example of a deterministic theory realizing that ideal. On the other hand, one can claim that processes which behave like Markov chains are actually deterministic systems with a high degree of instability. However that statement cannot be verified. Negative results of the abstract theory of dynamical systems make the classical ideal of deterministic explanation doubtful. The theory has, however, many positive and valuable consequences as it helps us to understand in what way amazing changes of structure appear in complicated material systems. It is a subject too detailed and technical for this philosophical conference.

Institute of Philosophy and Sociology
Polish Academy of Science

NOTES

[1] J. Watkins, *Science and Skepticism* (Princeton, 1984), p. 226. [For Watkins' B^* see p. 136].

[2] Among many books devoted to the problem of motion stability we mention three: V.I. Arnold, A. Avez, *Ergodic Problems of Classical Mechanics* (New York, 1968); J. Moser, *Stable and Random Motion in Dynamical Systems* (Princeton, 1973); A.J. Lichtenberg and M.A. Lieberman, *Regular and Stochastic Motion* (New York, 1983).

[3] B. Misra, I. Prigogine, M. Courbage, 'From deterministic dynamics to probabilistic descriptions', *Physica* **98A**, 1–26, (1979).

JOHN WATKINS

SCIENTIFIC THEORYHOOD

Within science the main units of appraisal, in the eyes of many philosophers of science, myself included, are scientific theories, rather than the various hypotheses that make them up or the concepts that occur in them or the research programmes within which they arise. And the problem on which I will focus is: what are the criteria for scientific theoryhood? More specifically: what conditions should a set of premises satisfy to constitute an axiomatization of *one* unified scientific theory, rather than be either a mere heterogeneous conjunction that does not amount to a theory, or else a collection of two or more disjoint scientific theories? To put the problem another way: it is sometimes said that scientific progress consists in subsuming more and more under less and less, which we may construe as meaning that more empirical phenomena come to be explained by fewer theories. But how should we understand this idea of a later scientific corpus containing *fewer* theories, with *more* empirical content, than an earlier corpus?

I call this the problem of *scientific theoryhood*. Quite a few philosophers of science have declared it unsolved, perhaps insoluble. It is connected, for reasons that will soon emerge, with another challenging problem, which we may call the problem of *natural axiomatizations*. A given body of propositional content, if it is finitely axiomatizable at all, can always be axiomatized in all sorts of different ways, some of them seemingly "natural", others highly "unnatural". But what requirements should "natural" axiomatizations satisfy? This again is a problem that quite a few philosophers of science have declared unsolved and perhaps insoluble.[1] In this paper I will summarize the solutions to these two problems presented in my *Science and Scepticism* (1984).

How do the premises of a genuinely unified scientific theory differ from such a heterogeneous conjunction of premises as

A_1: All cats are mousers
A_2: All cows are herbivorous
A_3: All crows are black

which obviously does not constitute a unified scientific theory? The answer can't be in terms of logical relations between the axioms; the members of any axiom-set, whether they are the axioms of a scientific theory or of any other body of non-analytic propositional content, must of course be logically inde-

J. Misiek (ed.), The Problem of Rationality in Science and its Philosophy, 221–229.

pendent of one another; there must not be inconsistencies between them and if one were entailed by the others, it would not be an axiom but a theorem. Nor can the answer be in terms just of testability or falsifiability. The above ragbag "theory" is falsifiable—by the observation of a man-eating cow, for instance, or of an otherwise normal cat gazing indifferently at mice scampering in front of it—and each of its axioms contributes to its falsifiable content. We could instead have taken for our A_1, A_2 and A_3 three falsifiable scientific theories that are quite unrelated to one another, say one in neurology, another in astronomy, and a third in economics. Nor can the answer be just in terms of the axioms being collectively fertile, in the sense of having, when conjoined, consequences that are not consequences of any proper subset of them. For any conjunction of axioms will automatically be fertile in that sense; "All cats are mousers if and only if all cows and crows are black or herbivorous" is one of the consequences of the above conjunction of axioms that is not a consequence of any proper subset of them.

But perhaps the answer is in terms of the axioms having excess *testable* content when conjoined? Let T denote an axiom-system consisting of axioms $A_1, \ldots A_n$, and let T' denote some but not all of these, and T'' the remainder, so that T' and T'' constitute a partition of T. Then we might try saying that T is a unified scientific theory if and only if, for every partition of T into some T' and T'', the testable content of T is greater than that of the sum of the testable contents of T' and T''. For this would tell us that each A_i in $A_1, \ldots A_n$ is *organically fertile*, in the sense that its presence enhances the testable content of the axiom-system by more than just the testable content, if any, of A_i on its own. Indeed, it may happen that A_i has *no* testable content on its own, and yet enormously enhances the empirical content of T. This would be the case if T were a fleshed-out version of Newtonian Mechanics consisting of Newton's laws of motion and law of gravitation together with the various auxiliary assumptions needed to relate this core to the behaviour of planets, comets, tides, etc., and A_i were Newton's First Law.[2]

This proposal faces a serious difficulty; but before turning to that I will first make the idea of the *testable content* of a scientific theory a little more definite. It is important here to avoid double-counting. It might sound plausible to say that the testable content of T is the set of all the testable consequences of T; but that would involve double-counting. For T is likely to entail testable experimental laws, such as $\forall x(Fx \rightarrow Gx)$, which in turn entail singular predictive implications, or SPIs as I call them for short, such as $Fa \rightarrow Ga$, which are also testable; and it would obviously be double-counting to include both that law and these SPIs in the empirical content of T.

Popper equated the testable content of a theory with the class of its potential falsifiers (henceforth PFs).[3] A SPI is the negation of the corresponding PF; thus $Fa \wedge \sim Ga$ is a PF of $\forall x(Fx \rightarrow Gx)$, and the negation of this PF, namely $\sim(Fa \wedge \sim Ga)$, or $\sim Fa \vee Ga$, or $Fa \rightarrow Ga$, is the corresponding SPI. PFs and SPIs have a pebbly character; sticking two pebbles together does not create an additional pebble, and cobbling two PFs, or two SPIs, together does not normally create an additional PF, or SPI. I equate the testable content of an axiom-set T, denoted by $\mathrm{CT}(T)$, with the class of its SPIs. Thus $\mathrm{CT}(T) \neq \emptyset$ says that T is testable (the class of its SPIs is not empty), and $\mathrm{CT}(T_2) > \mathrm{CT}(T_1)$ says that T_2 is more testable than T_1. This can come about in three ways. (i) T_2 is testable but T_1 is not. (ii) T_1 and T_2 are both testable and the proper subset relation holds between their testable contents, or $\mathrm{CT}(T_1) \subset \mathrm{CT}(T_2)$ (for this relation to hold, it is neither necessary nor sufficient for T_2 to entail T_1). (iii) A more interesting possibility, where T_1 and T_2 are both testable, is that we have $\mathrm{CT}(T_1) \not\subset \mathrm{CT}(T_2)$ but T_1 has what I call an incongruent counterpart, say T_1^+, and we have $\mathrm{CT}(T_1^+) \subset \mathrm{CT}(T_2)$.[4] The idea of incongruent counterparthood was introduced to enable us to handle a kind of case that has been very important in the history of science, namely one where a new theory both goes beyond and revises the testable content of its predecessor(s). Incongruent counterparts differ in what they say (congruent counterparts are logically equivalent), but are isomorphic in the sense that every consequence of one has a (congruent or incongruent) counterpart among the consequences of the other, and vice versa. So any SPI of the one will have a counterpart among the consequences of the other which will be a SPI of the other, and vice versa. So their different testable contents are equal in amount. We may represent this by $\mathrm{CT}(T_1) \approx \mathrm{CT}(T_1^+)$. Suppose now that, while we have $\mathrm{CT}(T_1) \not\subset \mathrm{CT}(T_2)$ because T_2 revises the testable content of T_1, we also have $\mathrm{CT}(T_1^+) \subset \mathrm{CT}(T_2)$. Then the testable content of T_2 is larger than that of T_1^+ which is equal in amount to that of T_1; hence the testable content of T_2 is larger in amount than that of T_1, or $\mathrm{CT}(T_2) > \mathrm{CT}(T_1)$.

In its most general formulation, this measure of comparative testability says that T_2 is more testable than T_1, or $\mathrm{CT}(T_2) > \mathrm{CT}(T_1)$, if and only if there is a (congruent or incongruent) counterpart T_1^+ of T_1 such that every SPI of T_1^+ is a SPI of T_2 but not vice versa.

I now come to the serious difficulty that faces this line of attack on the problem of scientific theoryhood. The idea, it will be remembered, is that if T is an axiom-system of a genuine scientific theory, then however we may partition it into T' and T'', the organic fertility requirement, OFR for short, will always be satisfied, or

$$\mathrm{CT}(T) > \mathrm{CT}(T') \cup \mathrm{CT}(T'').$$

This says that the class of the SPIs of T is larger than the union of the class of the SPIs of T' with the class of the SPIs of T''; or the testable content of the whole axiom-system is greater than the sum of the testable contents of (any two of) its parts.

But this won't do as it stands. For there will be ways of so reaxiomatizing any axiom-system T that we have $CT(T) = CT(T') \cup CT(T'')$. Here is one way. Let $A_1, \ldots, A_n$ be an axiom-system for some powerful scientific theory in which theoretical concepts figure essentially. Conjoin these axioms into a single axiom, and call this T. Now form a Ramsey-sentence T_R of T in the usual way, replacing each theoretical concept wherever it occurs in T by an existentially quantified predicate variable. Finally, reaxiomatize the theory with just the two axioms T_R and $T_R \to T$. Since a Ramsey-sentence captures all the empirical content of a theory, we have, contrary to OFR, $CT(T) = CT(T_R) \cup CT(T_R \to T)$, where $CT(T_R) = CT(T)$ and $CT(T_R \to T) = \emptyset$).

Here is a more complicated variant of the above, which I owe to Clark Glymour.[5] Let the axioms of a theory T be

A_1: $\forall x(Fx \to Qx)$
A_2: $\forall x(Qx \to Rx)$
A_3: $\forall x(Rx \to Gx)$

where F and G are observational and Q and R are theoretical predicates. Now take the strongest theorem of these axioms that contains only observational predicates; in the present case, this would be $\forall x(Fx \to Gx)$. Call this T'. We now form T'' in a way that is essentially similar to the previous way, except that the antecedent and consequent clauses are both governed by the same quantifier, and would, on their own, constitute open formulas rather than propositional units:

$$T'':\forall x((Fx \to Gx) \to ((Fx \to Qx) \wedge (Qx \to Rx) \wedge (Rx \to Gx))).$$

It is easily seen that the conjunction of the present T' and T'' is equivalent to T; and we again have $CT(T) = CT(T')$, contrary to OFR.

The problem then became to devise rules which would exclude all such OFR-defeating reaxiomatizations of scientific theories. We sometimes feel intuitively that one axiomatization is "natural" while an alternative is "unnatural"; and I hoped that these rules would disallow only "unnatural" ones. Instead of rehearsing here the rules I presented in the book,[6] I will say something about the motivation behind them.

As I mentioned, each axiom in a (non-analytic) axiom-system must be logically independent of all the others; so we need an *independence* requirement. Another requirement is indicated by the fact that one feels that a "natural" axiomatization will not amalgamate into one composite axiom what could have been presented as several relatively simple axioms. So we need a *decomposition* requirement. But decomposition can be carried too far and turn degenerate; instead of resulting in ever simpler units, it may in time lead to unnecessary complexity. This would happen if, after decomposing one complex axiom $A_1 \wedge A_2$ into the separate axioms A_1 and A_2, it went on to decompose A_1 into $A_1 \vee p$ and $A_1 \vee \sim p$. So an additional function for the other requirements is to put suitable constraints on the decomposition requirement. (But see the Postscript *below*.) Another rather obvious requirement on a "natural" axiom-system is that it should not contain predicates, constants, clauses, etc. that occur inessentially in the sense that they could be dropped without affecting the content of the system. We need a *nonredundancy* requirement.

I turn now to a more interesting requirement. For reasons that will emerge, I call it *Wajsberg's requirement*. In the first of the two OFR-defeating examples presented above, one axiom (namely T_R) recurs as a component of the other axiom (namely $T_R \rightarrow T$). And my earliest version of what was to become Wajsberg's requirement laid down that no axiom may recur as a component of another axiom. That was okay so far as it went, but it did not go far enough. For instance, it would not exclude the above-mentioned decomposition of A_1 into $A_1 \vee p$ and $A_1 \vee \sim p$, for there is no recurrence of an *axiom* in the latter. (Since p occurs inessentially in it, this "unnatural" decomposition would actually be excluded by the nonredundancy requirement; but there are other OFR-defeating reaxiomatizations that would not be excluded by either the nonredundancy requirement or my requirement that no axiom may recur as a component of another axiom.) My colleague Elie Zahar then suggested that I strengthen my requirement to say that an axiom is impermissible if it contains, as a (proper) component, a consequence of the axiom-system. He gave the following rationale for this requirement. Let c be a consequence of an axiom-system T that occurs as a component of an axiom in T. Then we will not increase the content of T if, as a first step, we give c the status of an axiom. And having done that, we will not decrease the content of T if we replace c wherever it had previously occurred within T by a tautology. We now comb the thus revised axiom-system, first with the nonredundancy requirement, eliminating all redundant tautological clauses, and then with the independence requirement, demoting to the status of theorem any "axiom" entailed by the other axioms. (This may, but need not, result in c being demoted.)

By way of a trivially simple illustration, let us try this out on our "unnatural" axiom-system, $A_1 \vee p$ and $A_1 \vee \sim p$. Since A_1 is a consequence of this axiom-system, we state it as an independent axiom, replacing it by the tautology t wherever it had occurred. This yields the three "axioms" A_1, $t \vee p$, and $t \vee \sim p$. After eliminating tautologies, we have just A_1.

When David Miller read a draft version of this, he pointed out that Zahar had been partially anticipated by the Polish logician Mordchaj Wajsberg (born 1902, perished during World War Two).[7] Miller had pieced together, from separate reports, Wajsberg's definition of an axiom as "*organic with respect to a system A* if it contains no segment which is a theorem of the system A 'or becomes one as soon as its variables have been bound by an appropriate quantifier'".[8] This suggested strengthening Zahar's requirement by adding that an axiom A_i is impermissible if it contains, as a proper component, a sub-clause that would be a consequence of the the axiom-system if its variables were bound by the quantifiers that bind them in A_i. This strengthened requirement, the *Wajsberg requirement* as I called it, will exclude the second OFR-defeating reaxiomatization presented above, due to Glymour. Here the second axiom, namely

$$T'': \forall x\, ((Fx \rightarrow Gx) \rightarrow ((Fx \rightarrow Qx) \wedge (Qx \rightarrow Rx) \wedge (Rx \rightarrow Gx)))$$

contains various sub-clauses, including (*F*x ® *G*x), that become consequences of *T¢* Ù *T*² when their variable is bound by the quantifier, namely "*x*, that binds it in the axiom.

I now say that a set of premises constitutes a unified scientific theory if it is logically equivalent to an axiomatization T that is "natural" in that it complies with the above requirements, and which cannot be partitioned into a T' and T'' such that $\mathrm{CT}(T) = \mathrm{CT}(T') \cup \mathrm{CT}(T'')$, or such that for every T' and T'' we have $\mathrm{CT}(T) > \mathrm{CT}(T'') \cup \mathrm{CT}(T'')$.

This criterion for scientific theoryhood helps to dispose of several difficulties that have plagued recent philosophy of science. One of these may be called the problem of the expanding/contracting beneficiary of corroborations:[9] when a novel scientific prediction is borne out by the result of a severe test, just what gets corroborated (confirmed)? The problem may be illustrated by a fable. Ms. Goody and Mr. Baddy are respectively authors of a powerful scientific theory, call it T, and a poor metaphysical hypothesis, call it M. Ms. Goody has derived from her T a challenging and novel predictive consequence c which has now been tested for the first time, passing with flying colours. Mr. Baddy now craftily asks Ms. Goody just what she considers to have been corroborated by this successful outcome. She answers: the theory by which the prediction was entailed, namely her theory T. He points out that the prediction was also entailed by $T \wedge M$, a powerful theory co-authored by him; the test result strikingly corroborates this Goody-Baddy Hypothesis, as he proposes to call it. She vigorously repudiates this claim, pointing out that, unlike his *M*, all the premises of her *T* were needed for the derivation of *c*: omit any one of them and the derivation no longer goes through.

Mr. Baddy now changes tack: if she is going to use that line of argument against $T \wedge M$ being the beneficiary of the corroboration, he will use the same line of argument against *T* being the beneficiary. He concedes that, as currently axiomatized by her, all the premises of *T* were needed for the derivation of *c*, while *M* was not; but on other axiomatizations the matter stands very differently. For instance, we could let the axioms be $\{M; M \rightarrow c; c \rightarrow T\}$. On this re-axiomatization, his *M* is indispensable: omit it and the derivation of *c* no longer goes through.

At this juncture a knowledge of Wajsberg's requirement will stand Ms. Goody in good stead, allowing her to point out that this Baddy reaxiomatization is impermissible: its second and third axioms both contain two components each of which is a consequence of the axiom-set. More generally, once we have a criterion for scientific theoryhood, we can dispel the problem of the expanding/contracting beneficiary of corroborations by laying down that corroborations can be won only by unified scientific theories, and not by ragbag collections or systems consisting of a scientific theory with bits tacked onto it.

In Watkins (1984) I introduced a further requirement that relies on a distinction between a theory's fundamental or core assumptions, and its subsidiary or auxiliary assumptions. Let *T* be a unified theory axiomatized "naturally" in accordance with our requirements, and denote by *S* those of its axioms that are subsidiary in some specified sense. Then I say that the corroborable content of *T* consists of CT(*T*) minus CT(*S*) or the SPIs of *T* that are not entailed by its subsidiary assumptions alone. In short, *T*'s fundamental assumptions have got to be involved in the derivation of a prediction *c* for *T* to gain a corroboration if *c* survives a test. This provides an answer to the following question. Suppose that evidence *e* which had refuted theory T_1 is now predicted (or retrodicted) by a new theory T_2: under what conditions does *e* corroborate T_2? The Zahar-Worrall view is that *e* must not have been used in the construction of T_2.[10] I say (i) that T_2's retrodiction, or explanation, of *e* must belong to its corroborable content, which means that its fundamental assumptions were essentially implicated, and (ii) that its corroborable content must be greater than T_1's. The satisfaction of (i) and (ii) would mean that T_2's ability to explain *e* is not ad hoc but is connected with the greater organic fertility of its fundamental assumptions.[11]

Another problem over which OFR can help us is that of so-called "Kuhn-loss". Let T_1 be an earlier theory superseded by the later T_2; and assume that T_1

and T_2 are both unified theories and that we have $CT(T_1) < CT(T_2)$. A proponent of the Kuhn-loss thesis might claim that the comparison has been rigged: T_2 should have been compared, not with T_1, but with some larger corpus, say C, from which T_1 has been artfully extracted to create an illusion of empirical gain. Well, that's a possibility; on the other hand, it might be that the proponent of the Kuhn-loss thesis has artfully expanded T_1 into C in order to create the illusion of Kuhn-loss. OFR could help us to tell which way round it is. If it turns out that C is a mere amalgam and that T_1 is the largest component of C to constitute a unified theory that is in competition with T_2, then there has not been Kuhn-loss.

In conclusion, I will offer a positive answer to a question posed earlier, namely whether a sharp meaning can be given to the idea that *more* empirical facts are now explained by *fewer* theories than at an earlier time, so that there has been a degree of theoretical unification. Assume: (i) that G, K and N are axiom-systems each of which satisfies our criterion for scientific theoryhood; (ii) that N came after G and K; and (iii) that the following two relations hold:

$$(1) \qquad CT(G \wedge K) = CT(G) \cup CT(K)$$

$$(2) \qquad CT(G \wedge K) \leq CT(N).$$

(I use the letters G, K and N because a paradigm of this pattern is the supersession of Galileo's and Kepler's laws by Newtonian Mechanics.) (1) tells us that the two theories G and K do not, when conjoined, combine to form one scientific theory, and (2) tells us that N has at least as much testable content as the conjunction of G and K. So *one* theory is now doing at least as much empirical work as had previously been done by two theories; there has been theoretical unification.

But my main claim for the Organic Fertility Requirement and for the rules for natural axiomatizations on which it depends, is that they assist those of us who place *theories* at the centre of our philosophy of science to say what theories are.

POSTSCRIPT, 1992

The above is the text, with only stylistic changes, of a talk given at the conference on "The Aim and Rationality of Science" at the Jagiellonian University, Krakow, in June 1989. Soon afterwards there appeared a powerful criticism of my rules for natural axiomatization, especially Wajsberg's requirement in combination with the Decomposition requirement, in Graham Oddie (1989). A replacement for Wajsberg's requirement that takes account of Oddie's objections is proposed in Zahar (1991). This Wajsberg-Zahar or W-Z requirement, as I call it, Watkins (1991), renders the Decomposition requirement redundant.

The London School of Economics

NOTES

[1] For references see Watkins (1984), pp. 204–206.
[2] For examples of the disasters that can ensue when the distinction between such a fleshed out theory and its core assumptions is overlooked, see Watkins (1984) § 8.52.
[3] See Popper (1959), chapter iv.
[4] For an explication of incongruent counterparthood, see Watkins (1984), pp. 171f.
[5] For fuller details, see Watkins (1984), pp. 209–212.
[6] Watkins (1984), pp. 208–209.
[7] See the editor's introduction to Wajsberg (1977).
[8] Miller (1974), p. 187.
[9] This problem is presented in Glymour (1980), pp. 30f; and see Worrall (1982), pp. 112–113.
[10] See Zahar (1973), pp. 101–104; and Worrall (1978), pp. 48–49.
[11] See Watkins (1984), § 8.26.

REFERENCES

Glymour, Clark, *Theory and Evidence*, Princeton: University Press, 1980.

Miller, David W., "On the Comparison of False Theories by their Bases", *The British Journal for the Philosophy of Science*, 25, 1974, pp. 178–188.

Oddie, Graham, "The Unity of Theories", in D'Agostino and Jarvie (eds.), *Freedom and Rationality: Essays in Honor of John Watkins*, Dordrecht: Kluwer, 1989, pp. 343–368.

Popper, Karl R., *The Logic of Scientific Discovery*, London: Hutchinson, 1959.

Wajsberg, Mordchaj, *Logical Works* (edited by S.J. Surma), Warsaw: Polish Academy of Sciences, 1977.

Watkins, John, *Science and Scepticism*, Princeton: University Press, and London: Hutchinson, 1984.

Watkins, John, "Scientific Rationality and the Problem of Induction: Responses to Criticisms", *The British Journal for the Philosophy of Science*, 42, September 1991, pp. 343–368.

Worrall, John, "The Ways in Which the Methodology of Scientific Research Programmes Improves on Popper's Methodology", in Radnitzky and Andersson (eds.), *Progress and Rationality in Science*, Dordrecht: Reidel, 1978, pp. 45–70.

Worrall, John, Review Article of Glymour (1980), *Erkenntnis*, 18, 1982, pp. 105–130.

Zahar, Elie, "Why Did Einstein's Programme supersede Lorentz's?", *The British Journal for the Philosophy of Science*, 24, 1973, pp. 95–123, 223–262.

Zahar, Elie, "Natural Axiomatization: a Revision of 'Wajsberg's Requirement'", *The British Journal for the Philosophy of Science*, 42, September 1991, pp. 391–396.

MIROSŁAW ZABIEROWSKI

ON THE OBJECTIVITY OF THE POPPERIAN INTERPRETATION OF QUANTUM MECHANICS

1. INTRODUCTION

Popper is one of the most ardent adherents of epistemology without a knowing subject.[1] His conception of objective knowledge has been expressed concretely in his interpretation of quantum mechanics devoid of an observer. We shall here consider his interpretation of quantum mechanics as founded on the *propensity* attribute of physical objects, and show that the Popperian exorcism of the observer from quantum mechanics does not have the intended effect. Thus, the Popperian *propensity* interpretation is neither subjectivist nor objectivist in the sense assumed by Popper. From our point of view the *propensity* interpretation appears to be anthropic.[2]

By the name 'anthropism', we refer to the trend in contemporary natural sciences which calls in question the extreme dualism of 'object-subject' and tries to formulate the criteria of the object-subject relation in an objective way. The Weizsaecker and Everett interpretations of quantum mechanics, Prigogine's thermodynamics, Mandelbrot's cosmography, are examples of such an anthropic approach.[3]

Before we start to examine the *propensity* interpretation we should firstly note the Popperian meaning of the observer's (experimentalist's) role in physics. Popper claims that quantum mechanics does not entitle us to consider the observer in a way different from classical physics because: "His task is to test the theory".[4] However, such a thesis cannot exclude the testing of theories which concern just the observer, or the process which an observer participates in. So the Popperian thesis of the role of the observer in physics is not conclusive indeed.

2. THE MEANING OF SUBJECTIVISM IN THE POPPERIAN EPISTEMOLOGY

The content of subjectivism is expressed by Popper as an 'interaction' between object and subject (measuring instrument). He does not give any explicit meaning to the term 'interaction'[5] which is essential for the discussion of

J. Misiek (ed.), The Problem of Rationality in Science and its Philosophy, 231–237.

the object-subject relation. Taking into account Popper's further considerations, it turns out to be evident that he does not understand the 'interaction' as Everett does; namely Everett's interpretation is judged by Popper to be fully objectivist.[6] Popper is not strict however when he maintains that the measuring apparatus in Everett's conception presents itself as a quantum system which is of the same kind as a quantum system of particles. In fact according to Everett only the system 'particle-measuring apparatus' is an object of quantum description.[7] From Popper's considerations we can deduce that he uses the term 'interaction' as an introduction of large changes to the measured system as an effect of the intervention of a measuring apparatus.[8]

It is not an accident that the meaning of subjectivity is not easily displayed from Popper's book on quantum mechanics because Popper assumes somewhat a fuzzy limit between the objectivity and the subjectivity of the measurement process: he considers degrees of objectivity and, in consequence, degrees of subjectivity. He regards the increase of objectivity as a result of the increase of measuring process indirectness.[9] According to him, a personal observer's participation in measurement is a source of the subjective influence upon the result of measurement although he claims the setting up of an experiment is not related to any observer's decision but to the objective theory which underlies every scientific experiment. In this place, Popper falls into contradiction: on the one hand objectivity can be graded, and on the other objectivity of a theory cannot. Popper tries to save the consistency of this reasoning when he argues that the objectivity of a theory is ultimately implied by its simplicity, symmetries, explanatory power, susceptibility to testing, etc. These criteria of objectivity are not, however, conclusive for distinguishing theories without an observer from those concerning processes which observers participate in.

3. 'LACK OF KNOWLEDGE' AS AN INTERPRETATION OF CLASSICAL AND QUANTUM PROPERTIES

Large parts of Popper's book on quantum mechanics impress the reader with its traditional style of discussion about the role of the observer in quantum mechanics. His argumentation does not take into account progress in natural sciences as well as in mathematics. For example the problem of the statistical interpretation of quantum mechanics presents itself in a wider realm than it was in the thirties. The development of so-called empirical mathematics and theoretical models of chaos throws new light on this question.[10]

We shall now consider whether the interpretation of probability as 'lack of knowledge' is peculiar only to quantum mechanics or also for other branches of physics. The second question which will be discussed is the status of statistical description in contemporary natural sciences.

Let us consider the three – body problem in classical mechanics. Similarly to two bodies interacting gravitationally, the appropriate dynamical equations can be easily formulated but in this case however a strict 'smooth' solution of them is not available. The solution of the three – body problem is characterized by instability points: some minimal perturbations of equation parameters (or of boundary conditions) can cause unpredictable effects, trajectory changes in the neighbourhood of these points. No fully deterministic solutions exist. In the cases of a larger number of interacting bodies, the solution of a non-linear system of equations is characterized by many instability points. Besides these points, the dynamical system has regular behaviour and it is generally possible to predict its irregularities (chaotic behaviour). However if the instability points are very numerous (or in the limiting case they form a Cantor set) it is impossible to find any kind of regular behaviour.

Popper connects the term 'lack of knowledge' exclusively with the subjectivist interpretation of probability. It is not always legitimate: in the cited non-linear dynamical systems, 'lack of knowledge' need not to be related to a limitation of the observer's senses or of the finite resolution of instruments (subjective factors). Moreover the interpretation of probability as 'lack of knowledge' need not be identified with quantum physics: in the case of classical non-linear system such an interpretation is also admitted[11] although here it is not subjectivist. 'Lack of knowledge' is here relativized to the knowledge available from the smooth solutions of dynamical equations but not to beliefs or cognitive abilities of the subject. 'Lack of knowledge' is here determined objectively by means of the instability points and the values of equation parameters (or boundary conditions). On the one hand, the statistical description is induced by practical necessity but on the other it reflects the true nature of the physical object (a non-linear system with many instability points).

Poppers is not right when he considers 'lack of knowledge' exclusively as a subjectivist factor. Of course, there is the problem of the subjectivist interpretation of the quantum wave function, in which the quantum wave function is to be the bearer of knowledge of the quantum system. Elsewhere[12] it has been shown that even in the latter case it is possible to reinterpret this 'consciousness' characteristic in an objective way.

An example of such an objectivation is Weizsaecker's interpretation of quantum mechanics which Popper does not discuss.

4. THE POPPERIAN *PROPENSITY* INTERPRETATION OF QUANTUM MECHANICS

Popper derives his *propensity* interpretation of quantum mechanics from the classical interpretation of probability.[13] The classical interpretation is applied in those cases where there independent possibilities appear (excluding each other) with identical chances. The term 'identical' chances should be just defined. The definitions received so far are subjectivist: 'chance' is an attribute of events or objects and there are no ways to define 'identical chances' without referring to the knowledge of the observer who uses the principle of insufficient reason.

Popper introduces the *propensity* (of an event or object) to realize alternative behaviour or states. Similarly to the case of chance, it is not possible to relate the *propensity* exclusively with a 'naked' event or a 'naked' object, without referring to the 'physical situation'.[14] When Popper explains the sense of his interpretation and turns to prove its objective character, he uses arguments quite similar to those used by Weizsaecker[15]: he considers possible (virtual) experiments which are described by probabilities and actual (realized) experiments which are described by statistical frequencies. Contrary to Popper's discussion, Weizsaecker tries to develop the consequences of such assumptions: the actualized possibilities (n-alternatives) not only provide information about the object but also contain knowledge meant as a factor participating in quantum events.[16] Popper, though a philosopher, also pays no attention to clarifying the ontological status of his potentialities and actualities.

Popper ultimately claims that the *propensity* is not attached to a particle itself but to an objective physical situation in which the particle participates. The act of measurement can arrange such a situation. It implies the connection of the statistical interpretation with the result of the measurement process but not with the particle's property. For in the measurement process participate both the particle and the observer Feyerabend, Jammer, and others[17] impute subjectivism to the Popperian *propensity* interpretation. Popper defends the objectivity of his interpretation, arguing that *propensities* are associated with the physical situation which may be empirically controlled. Thus he identifies the objectivity with the managebility to checking. However if he assumes the controlling to reveal objectivity, he cannot wink at the participation of an observer in the measurement process, i.e. a kind of 'interaction' between subject and object.

5. THE ANTHROPISM OF THE *PROPENSITY* INTERPRETATION OF QUANTUM MECHANICS

The troubles of Popper, an avowed objectivist suspected of subjectivism, are an excellent illustration of the inadequacy of a 'two-valued' epistemology a 'objective-subjective'. Referring his *propensity* interpretation to the measurement situation, he chooses a third outcome indeed. He does not call it so but we define it as anthropic. Here are the anthropic elements of his interpretation:

1. The participation of the subject in the measuring process is a result of the necessity to actualize the virtual experiment, to create the indispensable physical situation. Like as the Weizsaecker's interpretation, the propensity interpretation is not subjectivistic because it refers neither to any individual features of the observer nor to his beliefs.
2. The very concept of *propensity* is anthropic. The *propensity* cannot be related to the classical object described by deterministic equations and abstracted from its surrounding. The *propensity* (an anthropomorphic term) is associated with the physical object understood in Prigogine's sense.[18] It not only exists but is becoming. It resembles rather a biological or self-organizing structure than a 'dead' naked object of classical physics. The *propensity* is associated with the process: in the process of becoming the *propensities* are realized with respective frequencies. If the *propensities* or tendencies are considered, an aim or a function of process must be assumed.
3. If the process consists in becoming, if its history can be realized in many but regular ways, then the process should be characterized by a kind of ability to 'recognize' its own present and history, and to 'project' his own future. All these attributes are not subjectivist but anthropic. Almost all processes proceed along with the man's consciousness. The point is not an absurd attempt to transfer the observer's consciousness into the physical object investigated. The point is that every process which is irreversible in time, has its own individual history and future, and it needs to have a kind of self-knowledge to be able to develop in time and to self-organize. In many cases, the self-knowledge can be objectively recorded and the process can be controlled.

6. GENERAL CONCLUSIONS

Here are the general conclusions of our considerations:

1. In the light of contemporary natural sciences, it is no longer possible to

keep a sharp opposition between subjectivism and objectivism with respect to cognition of physical processes. This opposition causes disputes which often result from the anthropomorphic terms which bring an easy attack of objectivists against a supposed subjectivism. Therefore we propose a third outcome, namely an anthropism which accounts for the objective recording of the subject-object relation both for the case when the subject is a real participator of the process and for the case when the process is considered as simultaneously subject and object of becoming.

2. The *propensity* interpretation is neither objectivist in the traditional sense nor subjectivist as the opponents of Popper wish. It is anthropic, as the Weizsaecker's and Everett's quantum interpretations are.
3. The category of self-knowledge need not be exclusively subjectivist and thus connected with human beliefs, wills etc.The anthropic approach, is understood as an ability to recognize oneself in one's own situation and can be an attribute of processes in which a man does not participate.
4. It is necessary to perform a separate extended study of the meaning of realism in anthropic physics.

Wrocław Technical University

NOTES

[1] K. R. Popper, *Objective Knowledge*, Clarendon Press, Oxford, 1981; *Unended Quest*, Fontana-Collins, Glasgow, 1982(a); *Quantum Theory and the Schism in Physics*, Rowman and Littlefield, Totowa, New Jersey, 1982(b).

[2] The name 'anthropic' is linguistically derived from the term 'anthropic principle' formulated in cosmology in the sixties and seventies. Our understanding of the word 'anthropic' is more extended. It is correlated with its cosmological counterpart only in this sense, that the existence of human beings as a part of nature should be taken into account in the physical description of nature. Almost all philosophical discussion related to so-called strong anthropic principle is of no influence on our understanding of anthropism. Here anthropism means the standpoint in epistemology which (a) admits the objective description of processes in which the observer participates; (b) investigates the relation between subject and object in scientific cognition; (c) searches for an objective expression of anthropomorphic concepts in natural sciences. Cf. J. D. Barrow, F. J. Tipler, *The Anthropic Cosmological Principle*, University Press, Oxford, 1986; C. W. Davies, *Prog. Part. Nucl. Phys.* **10** (1983) 1; M. Zabierowski, "The Anti-Newtonian Concept of the Observer", in *Isaac Newton's 'Philosophiae Naturalis Principia Mathematica'*, ed. W. A. Kamiński, World Scientific, Singapore, 1988(a), p. 178; "The Weizsaecker's and Everett's Interpretations of Quantum Mechanics in the Light of Anthropism", in *Problems in Quantum Physics*, ed. L. Kostro *et al.*, World Scientific, Singapore, 1988(b), p. 880; *Astrophysics and Space Science* **141**, 1988(c), 333; **201**, 1993(a), 125; **202**, 1993(b), 173; *Status of the observer in modern physics*, (in Polish), Institute of Physics, Wrocław 1990; "The Universe and the Man", (in Polish), Wrocław Technical University, Wrocław 1993(c); "Configurations and

Redshifts of Galaxies", in *Frontiers of Fundamental Physics*, ed. M. Barone and F. Selleri, Plenum Publishing Corporation, New York/London/Washington/Boston, 1994(a), p. 37; "Fornax – the companion of the Milky Way and the Question of its Standard Motion", in *Problems of Fundamental Physics*, ed. F. Selleri and M. Barone, Plenum Publishing Corporation, New York/London/Washington/Boston, 1994(b).

[3] Cf. C. F. von Weizsaecker, *The Unity of Nature*, Farrar, Straus, and Giroux, Inc., New York, 1980; H. Everett, *Rev. Mod. Phys.* **29** (1957) 454; I. Prigogine, *From Being to Becoming*, W. H. Freeman Co., San Francisco; 1980; I. Prigogine, I. Stengers, *Order out of Chaos – Man's New Dialogue with Nature*, Heinemann, London, 1984; B. B. Mandelbrot, *Fractals, Forms, Chance, and Dimension*, W. H. Freeman Co., San Francisco, 1977; Zabierowski, 1988 (abc), 1990, *op. cit.* in ref. 2, *Zagadnienia Naukoznawstwa* **1–2** (101–102), (1990) 163.

[4] Popper, 1982(b), *op. cit.* in ref. 1, p. 35.

[5] *Ibid.*, p. 40.

[6] *Ibid.*, p. 90.

[7] Everett, 1957, *op. cit.* in ref. 3; Zabierowski, 1988(b), *op. cit.* in ref. 2; T. Grabińska, *Ronnitu Filozoficzne* XXXVII–XXXVIII, z. 3, (1989–1900), 53.

[8] Popper, 1982(b), *op. cit.* in ref. 1, p. 41.

[9] *Ibid.*

[10] Mandelbrot, 1977, *op. cit.* in ref. 3; T. Grabińska, *Elektronik* **4** (1986) 3, (in Polish); *Zagadnienia Naukoznawstwa* **XXIII**, (1987) 329 (in Polish); Prigogine, Stengers, *op. cit.* in ref. 3; T. Grabińska, *Realism and Instrumentalism in Modern Physics*, Wrocław Technical University, Wrocław 1992.

[11] Popper, 1982(b), *op. cit.* in ref. 1, p. 50.

[12] Weizsaecker, 1980, *op. cit.* in ref. 3; Zabierowski, 1988(b), *op. cit.* in ref. 2.

[13] Popper, 1982(b), *op. cit.* in ref. 1, p. 68.

[14] Popper, 1982(b), *op. cit.* in ref. 1, pp. 69–72.

[15] C. F. von Weizsaecker, "A Reconstruction of Quantum Theory", in *Quantum Theory and the Structures of Time and Space*, vol. 3, ed. L. Castell and C. F. Weizsaecker, Carl Hanser Verlag, Muenchen, 1979; *op. cit.* in ref. 3.

[16] The understanding of knowledge as a factor participating in quantum events is discussed in Weizsaecker, 1980, *op. cit.* in ref. 3 and Zabierowski, 1988(b), 1990, *op. cit.* in ref. 2.

[17] Cf. P. K. Feyerabend, *Philosophy of Science* **35** (1968), 309; *Philosophy of Science* **36** (1969) 82; M. Jammer, *The Philosophy of Quantum Mechanics*, John Wiley, New York, 1974.

[18] Prigogine, 1980, *op. cit.* in ref. 3; Prigogine, Stengers, 1984, *op. cit.* in ref. 3; T. Grabińska, *Theory, Model, Reality*, Wrocław Technical University, Wrocław 1993; M. Zabierowski, *Anthropic Cosmology*, Wrocław Technical University, Wrocław 1994 (c).

JÓZEF ŻYCIŃSKI

HOW TO DE-RUSE SOCIOBIOLOGICAL THEORY OF KNOWLEDGE?*

SCIENCE IN EVOLUTIONARY PERSPECTIVE

The epistemological significance of E.O. Wilson's sociobiology has been enthusiastically proclaimed by Michael Ruse in his version of the so-called Darwinian epistemology. When trying to develop its tenets, Ruse argues that the principle of the survival of the fittest should be consistently applied not only to the domain of biological organisms but also to scientific theories and their competition in the struggle for survival. This claim seems scarcely original since in authors as remote as Herbert Spencer, Ernst Mach, and Karl Popper we find its anticipations in various metaphors that describe the growth of scientific theories in biological terms related to natural selection. Ruse's declarations seem, however, much stronger than the suggestions mentioned above when he claims that 'there are very good biological reasons' to regard mathematical theorems as objective truths because the theorems at stake give us 'a selective advantage' in the struggle for survival.[1] Such a statement must not result in the rejection of epistemological realism or imply an irrationalist stance in the theory of knowledge. Ruse explicitly declares that as a former logical empiricist he was and still is 'attracted to the rationality of science', and still asserts that 'the course of science is . . . not totally without sane reason.'[2]

In this paper I try to answer, what precisely does the mysterious 'attraction' mean and how can a critical version of sociobiological epistemology be reconciled with the standpoint of metascientific rationalism. My answer to the latter question is that sociobiology and rationalism can be reconciled only if one rejects, or substantially modifies, those statements in Ruse's expositions that seem much closer to rhetoric than to critical metascientific reflection.

ARE SCIENTIFIC THEORIES ADVANTAGE-LADEN?

When developing his version of a sociobiological theory of knowledge, Ruse presents it as a natural consequence of the Darwinian approach and argues that 'a century and a quarter after the appearance of *On the Origin of Species*, the time has surely come to take Darwin seriously.'[3] This very statement suggests

J. Misiek (ed.), The Problem of Rationality in Science and its Philosophy, 239–252.

Kluwer Academic Publishers.

that for 125 years Darwin was not treated seriously, and that to be a serious Darwinist one must introduce to the theory of knowledge the strong assumptions accepted by E.O. Wilson in his radical version of sociobiology.[4] The question arises, however, whether at the same time, in the evolutionary epistemology, one can treat seriously the rational heritage of mankind and its particular role in our culture.

To answer this question we must assess two assumptions substantial for the sociobiological theory of knowledge. They assert respectively that:

1. The content of scientific theories is biology-laden as well as genetically conditioned.
2. In the growth of human knowledge, the victorious are those theories that give evolutionary advantages in the struggle for survival to the human species.

These vague statements could imply essentially different epistemological stances depending on the meaning ascribed to such terms us 'conditioned,' 'victory,' 'advantage.' There is nothing *per se* controversial in Ruse's arguments when he claims that 'scientific methodology is grounded in epigenetic rules, brought into existence by natural selection.'[5] To determine the exact content of such statements one has to define, however, the precise meaning of the expression 'grounded in' and its metaphorical counterparts that are abundantly used by Ruse. It is hard to determine a clear meaning when Ruse aphoristically ascertains that our knowledge is 'biologically based' on these rules, and that epigenetic rules 'influence our thought.'[6] Depending on what meaning we ascribe to the terms 'grounded,' 'based,' 'constrains,' 'influences,' we may obtain either a trivial or an obviously irrational theory of knowledge.

Nobody would question Ruse's statement that 'the proto-human who innately preferred '2 + 2 = 4' to '2 + 2 = 5' was at a selective advantage over his/her less discriminating cousin.'[7] Nonetheless, in another text the same notion of the 'selective advantage' dependent on knowledge seems semantically less clear, when, for instance, Ruse contends that 'the human who instinctively recognizes and prefers sweet things to sour or rotten things is at a clear advantage to the human whose palate is indifferent to tastes.'[8] This contention seems at least controversial since one might argue as well that human indifference to palate pleasures would facilitate our struggle for survival because it makes us indifferent to and independent of the quality of the food available.

The semantic fuzziness of the basic notion of the 'selective advantage' is well illustrated by A. Zahavi's paradoxical conception of the selective advantage of having a handicap.[9] In this approach, a handicapped animal could increase his genetic successes if he attracts females by the very fact that in spite of his biological disadvantages, he is able to participate participate in the struggle for survival. In such an interpretive approach, dialectic is mixed with rhetoric; disadvantage means advantage, and any conclusion can be deduced from the set of ambiguous elementary assumptions.

Setting aside discussion of the precise meaning of the basic concepts that are used in sociobiological epistemology, we may restrict our attention to those simple cases that are evidently non-controversial. Such is the case with the thesis of a biological basis of human knowledge. Except for eccentric defenders of the theory of a *tabula rasa*, nobody would question either that our biology in a sense 'influences' and 'constrains' our mathematics or that, in accordance with the famous phrase of Wilson, the content of scientific theories also is, in a sense, 'held on a leash' by genes. If we had the brain of a macaque, we certainly could not develop the black hole physics and the quaternion calculus. Most important, however, the question remains how long is the genetic leash, and how strong is its impact on the content of scientific knowledge.

In his many statements, Ruse definitely rejects the trivial interpretation of his naturalized epistemology. He consistently accentuates the novel character of the Wilsonian approach and expresses his disappointment with both uncritical rationalism and the absence of Darwinian patterns in traditional epistemology. He is especially disappointed with the situation in the philosophy of biology where before his analysis of evolutionary theory in his PhD Thesis, 'the literature on the subject was limited, and much of it was very bad.'[10] Similar wording suggests that Ruse would not accept a trivial interpretation of his new metascientific patterns since he announces radical revisions in the traditional theory of knowledge.

There seems to be a model in which one might accept the two sociobiological assumptions mentioned above as well as the traditional version of the metascientific rationalism. I will call such an interpretive proposal the correspondence model. The term 'correspondence' indicates that one must not introduce a mutual opposition between the objective content of scientific theories and the biological basis of the thinking subject on the one hand, and the truth and evolutionary advantages on the other. It is, at least theoretically, possible that these three factors remain autonomous and consistent. It means that the content of scientific theories could be neither reduced to biological determinants nor explained by selective advantages despite the harmonious coexistence, respectively, of the biological, intellectual and evolutionary elements.

It is true that such a three-level correspondence seems close to a science-fiction epistemology in which biology itself generates truth which, like a good sheriff in an unsophisticated movie, always wins in the struggle for survival. Similar objections are, nonetheless, aesthetic in nature; the domain of artificial intelligence and the competition between computer programs provides at the same time an interesting example which could be useful in illustrating the basic epistemological tenets adopted in such a model, consistent with tenets of moderate sociobiology.

The information contained in computer software has its electromagnetic basis in the appropriate hardware. In a sense the content of the software is held on the leash by hardware, since primitive or wrecked computer models could impede the operation of complicated and/or sophisticated programs. When approaching the situation statistically, one could argue also that market successes of particular companies depend on the objective value of their products. If WordPerfect wins in competition with ChiWriter, it is probably due to its objective superiority defined in terms of utility for the average user. However, neither trade competition nor the electromagnetic basis of computer software determines the objective value of the output information content.

The truth of logical relations encoded in the software enters into a complex network of relationships dealing with both physical processes and market rules of competition. It would, however, be naive to argue that the content of software is entirely generated by electromagnetic impulses or could be explained by the very analysis of a competitive struggle for existence between various technological companies. Quite the contrary, experts in computer science regard as a commonplace the proposition asserting that 'electronics is almost totally irrelevant to the nature of artificial intelligence.'[11] Only a few general properties are required in computer mechanisms to obtain a system capable of effective information processing. After providing such basic features as a set of internal states, which makes elementary operations possible, as well as a symbolic memory accessible to the operations of writing, reading and addressing, technological details do not affect substantive results that depend on the logical structure of software. These results cannot be regarded as products of the hardware basis and the content of software instructions cannot be regarded as a reality constituted exclusively by electromagnetic impulses in the technology involved.

The example provided indicates that, without questioning the traditional metascientific rationalism, one could adopt the basic epistemological tenets of moderate sociobiology together with the thesis of the mutual correspondence between biological, rational, and evolutionary factors. This correspondence model would allow us to defend the notion of objective truth because the relationships between the heterogeneous elements in question would be truth preserving; one cannot, however, attempt to consider the objective content of scientific theories as a product either of genetic determinants or of evolutionary needs. Such a strong reduction, implying an arbitrary ontological monism, would remain clearly inconsistent with the facts well-known from the field of artificial intelligence, and will have no independent confirmations as well.

In contradistinction to the suggested model, in classical texts of radical sociobiologists we find strong statements suggesting that the objective content of scientific theories, religious beliefs and moral principles can be entirely explained by analysis of their biological basis. Articulating this version of reductionism, E.O. Wilson categorically insists in his classical work that 'the possibility of explaining traditional religion by the mechanistic models of evolutionary biology . . . will be crucial. If religion, including the dogmatic secular ideologies, can be . . . explained as a product of the brain's evolution, its power as an external source of morality will be gone forever. . .'[12] If the computer scientists argued in the same manner, they would plainly explain the nature of computer software by arguing: 'If the information processing is explained in terms of hardware's mechanisms, the content and results of the programme would be completely unessential and their informative power will be gone forever.'

The same biological reductionism appears in Wilson's explanations of the content of moral principles when he argues: 'if termites had somehow managed to ascend . . . to the pinnacle that human beings now occupy, . . . we would find them rationally explaining an ethical system with myths, legends, sacred literature, and a termite God profoundly different from our own. Their theologians would advance, as absolute moral guidelines, a sacred caste system, cannibalism glorified, personal reproduction a sin in the worker castes, territorial war proper, darkness preferred to light. . .'[13] In the context of this prophecy, Wilson mentions Robert Nozick's defence of the rationality of science presented in his book *Philosophical Explanation*. Nozick suggests there a possibility of the existence of certain objective relations and principles that human beings are tracking by biological and, in particular, genetic means. Nozick's interpretation remains consistent with the examples from the field of artificial intelligence provided above and accentuates as well the impossibility of reducing the content of human knowledge to purely biological factors. Wilson reacts to this interpretation with his usual rhetoric. He answers charmingly: 'If that much is true, we have a basis for extrabiological origins of moral reasoning. But I am not sure the termites would agree.'[14]

The theory of knowledge developed from the termites' point of view by the famous insect specialist finds a consistent extension in Ruse's evolutionary epistemology. The author of *The Darwinian Revolution* accentuates the role of illusions underlying scientific activity and provides examples of the naive and uncritical belief in reason. However, in his comments Ruse avoids biological reductionism and introduces important distinctions when determining the criterion of truth of particular theories. In his metaphoricalal language, 'science is attached to biology' because 'it has its feet in Darwinian forces. But its head reaches up into non-adaptive clouds, as we push our enquiries further and further.'[15]

Within this framework, basic scientific principles depend on a 'mind-injected element' determined by our genotype. These principles are, however, used by human beings to develop sophisticated scientific theories. Their content remains related to evolutionary goals of our species; nonetheless, it cannot be regarded as an effect of necessary genetic determinants. Consistently, in sociobiological epistemology one cannot reduce the content of scientific theories to the functions of our genes, but one must distinguish the diverse epistemological statuses of particular theories. All theories and theorems are biology-laden, but their informative value differs substantially, being dependent on many extra-genetic factors. To precisely determine these differences, one must raise the problem of the relationship between scientific realism and the sociobiological theory of knowledge.

MYTHOLOGY AND REALISM IN THE THEORY OF KNOWLEDGE

According to the classical explanation proposed by Wilson, any form of knowledge expresses a mythopoietic drive of humankind and, consequently, contains a version of scientific mythology where the term 'myth' is understood in a non-pejorative sense. The so-called scientific materialism, based on the principles of sociobiology, 'is accepted as the more powerful mythology' because it 'is the only mythology that can manufacture great goals from the sustained pursuit of pure knowledge.'[16] In the adopted terminological convention, mythopoietic components of knowledge do not exclude the possibility of accepting the standpoint of cognitive realism. Evolutionary epics intermingle with the realistic description of physical facts in what is called by Ruse 'common-sense realism.'[17] Determining the exact content of this epistemological standpoint seems scarcely possible because Ruse states his clear opposition to the scholastic and Popperian practice of introducing precise definitions and minute divisions.[18] Consequently, he restицts himself to repeating the Quinean dictum that science is nothing but self-conscious common sense, and formulates later a rather cautious opinion that chairs, tables, and trees exist in reality, and are not chimeras like Macbeth's dagger.[19] He is inclined to recognize also the existence of electrons, genes, and dinosaurs; the importance of his ontological commitment is, however, lessened by the fact that it is only our biological propensity that 'provides us with our criteria of truth and reason.'[20] Consequently, 'it is difficult to know, in human terms, precisely what nature a thing-in-itself would possess'[21] since our commitment to a particular ontology is nothing but a consequence of interaction between epigenetic rules and biological environments. The role of biologically conditioned epigenetic rules is such that 'we should not be surprised when this latter [i.e. a person-independent-J.Z.] reality collapses into paradox and non-being.'[22]

Such a general description of epistemological tenets can be consistent with such remote cognitive standpoints as critical realism and van Fraassen's constructive empiricism. On the one hand, we find in Rusean epistemology non-controversial statements that 'human thought is moulded and constrained by the epigenetic rules.'[23] On the other, the rational accomplishment of mankind seems to be called into question by him when he declares that 'vanity and ignorance alone support the claim that human reason has a privileged status.'[24] Whatever the 'privileged status' means, the importance of the latter statement is immediately softened by his ascertaining that our intellectual achievement has 'a meaning of its own, transcending biology.'[25]

The large spectrum of essentially different statements allows us to reconcile his sociobiological theory of knowledge with substantially different epistemological stances. To counteract this semantic fuzziness and to determine the exact content of the Rusean version of the common-sense epistemological realism, we may focus on his interpretation of mathematics. In this very discipline, the vague terms of mythological components and evolutionary advantages imply consequences that are particularly easy subjects for critical appraisal. The imprecise declarations about explanatory myths and the intellectual struggle for survival take on a new meaning when we confront them with the reality of mathematical knowledge. Mathematics, regarded both as a most original achievement of the human mind and as our non-observable culture, can be especially useful in answering the question, what role is played in sociobiological epistemology by the objective content of our knowledge.

EVOLUTIONARY METAMATHEMATICS

Admitting that developed philosophy of mathematics, Ruse specifically denies the traditional conception of objective mathematical truths and maintains that the apparent objectivity of mathematics is illusory.[26] The essence of the so-called Darwinian approach to mathematics is to be expressed in the thesis that human individuals themselves objectify the truths of logic and mathematics, because personal certainty gives us a selective advantage in our struggle for existence.

Probably nobody would question that knowledge of mathematics and logic is really useful for our species. Selective advantages, however, can result from knowing not only elementary arithmetic, but also trivial facts dealing with prosaic realities of everyday life. The question arises then: Is our knowledge useful because it is true, or rather is it regarded as true because it appears useful in the struggle for survival? The fact of the evolutionary utility of subjective belief in the truth of certain statements scarcely provides an ultimate criterion of truth. Subjective conviction of the certainty of mathematics may have evolutionary usefulness, but an analogous psychological feeling of certainty can be generated by ideological dogmas or pseudo-scientific speculations. A sociobiological philosophy of mathematics does not explain what difference there is between the objective certainty of mathematical theorems and the subjective certainty of supporters of Nazi anthropology or Lysenko's biology. It does not explain why in elementary arithmetic we have no mathematical Lysenkos who, preserving the standard meaning of the employed terms, would argue that '2 + 2 = 5'. Such nonstandard arithmetic would certainly provide evolutionary advantages for particular groups demonstrating their territorial or financial supremacy in the struggle for existence. If Lysenkos appear in biology rather than in mathematics, it seems to suggest that the very essence of mathematical knowledge differs from the essence of biological sciences. Neither Ruse nor any of his collaborators explains what constitutes the essence of this difference.

In the same style Ruse ignores the question as to what evolutionary advantages result from particular mathematical discoveries. One may provide many examples that led to immediate disadvantages and existential troubles. It would be enough to mention here the drama of Georg Cantor. To his theory of infinite sets a radical opposition of the institutional science arose represented by Leopold Kronecker. The conflict resulted in Cantor's psychic breakdown, severe depression and death in a mental hospital. His intellectual legacy was recognized in a short time by authors who together with David Hilbert did not want to leave the new 'mathematical paradise' discovered by Cantor. The discoverer himself, however, except for an intellectual advantage, gained no other evolutionary advantages. Contrary to sociobiological principles, important intellectual achievements yielded the definitive defeat in his struggle for survival.

Ruse himself distinguishes different statuses of various mathematical theorems. He seems to combine sociobiology and intuitionism when he argues that there is a set of epigenetically determined simple mathematical principles and rules which appear to our consciousness as self-evident. In another context he seems to express his support for a modified metamathematical formalism, when he argues that after adopting the elementary set of genetically determined truths, logicians and mathematicians demonstrate their creative fantasy in developing new systems of fantastic games. More advanced mathematics is just 'an epiphenomenon on a biologically based set of simple statements and rules.'[27]

This approach reminds us of Kronecker's thesis that the integers were made by God, and everything else is the work of man. One difference is that in Ruse's philosophy God is replaced by epigenetic rules, and another that nobody knows which mathematical axioms are determined by these rules. The optimistic faith that it is enough to have a set of 'simple statements and rules' to construct the entire body of mathematics turned out to be definitely too optimistic when, at the beginning of our century, basic divergences of opinions emerged in the quest for fundamental axioms to which mathematics was supposed to be reduced.

Without paying special attention to metamathematical discoveries of our century, Ruse tries to apply his scheme to explain the genesis of non-Euclidean geometries. He suggests that the Euclidean fifth postulate, though suspect, remained unquestioned till the 19th century because it was relatively unimportant in the evolutionary struggle for existence. When taking part in this struggle, our *Australopicine* ancestors had to 'know that a straight line is the quickest way from A to B. But who cares about whether or not parallel lines never meet?'[28] One may doubt whether or not the concept of parallel lines was indeed pragmatically useless for primitive societies but consistently one should ask why these societies were so interested in theoretical problems that deal, for instance, with the equality of all right angles. The fourth Euclidean postulates asserts that all right angles are equal to one another and this postulate aroused no objections though its pragmatic utility seems rather dubious and *Australopi:ine* rather should not be disturbed with its content.

Perhaps sociobiology may be used to explain the genesis of certain concepts in pre-Euclidean mathematics. Euclid's *Elements* with its basic notions of indivisible points, breadthless lines and infinite surfaces remain, however, as distant from evolutionary adaptive advantages as Wilson's evolutionary epistemology is distant from contemporary mathematics. The most important questions that attracted the attention of successors of Hilbert and Gödel are in this epistemology either ignored or left with metaphoricalal maxims. Ruse's epistemological comments on the significance of Gödel's incompleteness theorem may well illustrate this procedure. In clear inconsistency with Ruse's philosophy of mathematics, this theorem can be regarded neither as intuitively self-evident nor as an arbitrary convention. It is not self-evident because its discovery was a shock for logicians. It is not a convention resulting from creative play with symbols, because there exists no mathematical game in which our creativity could be expressed in defining a system which is simultaneously: (1) isomorphic with *Principia Mathematica*, (2) consistent, (3) complete.

The fact of the incompleteness of arithmetic, regardless of its possible significance for our evolutionary development, appears as independent of both our conventions and our intuitions; the notion of mathematical objectivity is especially conspicuous when we consider the amazing results of Gödel's discovery. When many argue that this discovery belongs to the greatest discoveries of humankind and the questions inspired by them 'are pregnant with possibilities and fraught with dangers,'[29] Ruse in his sociobiological epistemology merely notices that 'worries about failures in completeness are vestiges of Platonic/theistic thinking.'[30] This approach illustrates the well-known method in which difficult questions are avoided and replaced by a distribution of philosophical labels. The content of the disturbing theorem provides, nevertheless, an especially suggestive counterexample to the sociobiological philosophy of mathematics. Incompleteness of a logical system appears as an objective counterintuitive fact irrelevant for biological selection. The easiest procedure to eliminate questions of the objective nature of this incompleteness is to impute Platonic or religious influences to philosophical opponents.

Vague metaphors in which one ignores long debates in the foundations of mathematics may appear attractive for epistemological radicals searching after simple explanations of the nature of mathematics. Simplicity of their image of mathematics results, unfortunately, from the systematic ignoring of basic discoveries of our century that revealed essential flaws in simple metamathematical schemes. One cannot effectively eliminate these flaws by introducing ambiguous terms to define relations of primary importance. Their ambiguousness results in interpretive arbitrariness. For instance, Wilson's statements about the genetic leash can be reconciled with mutually inconsistent standpoints in the philosophy of mathematics. An attempt to interpret mathematics as an evolutionarily useful illusion ends in semantic chaos, as well as in producing new interpretive illusions shared at a metamathematical level.

EVOLUTIONARY ADVANTAGES AND INTERPRETIVE TROUBLES

In certain respects, the sociobiological theory of knowledge seems to resemble the famous philosophy of Panglossianism ridiculed by Voltaire in his critique of Leibniz. The Leibnizian Doctor Pangloss cultivated the strong conviction that we live in the best of all possible worlds. Unfortunately, except for his subjective optimism, he had no other arguments to justify his views. In sociobiological epistemology, one possesses no arguments that the evolutionary interest of genes constitutes the most important criterion of truth. The thesis is adopted axiomatically to obtain a simple philosophy in which former epistemological debates are instantaneously dissolved.

In the same manner sociobiology dissolves several problems that are elementary for contemporary science and its philosophy. Conjectural perhapses are adopted as unquestionable axioms to find an instant solution to the most complicated questions. While certain statements imply that sociobiology conclusively disclosed the ultimate biological roots of mathematics, other announcements are much more modest in style, and only suggest that the epigenetic rules 'probably play a key role when we think mathematically.'[31] This cautious opinion has its only justification in a common-sense remark suggesting that 'innately, we think in terms of symbol and quantity because such a way of thought proved its adaptive worth.'[32]

The common-sense mathematics regarded as a mixture of quantities and symbols has its intellectual counterpart in naive anthropomorphic physics in which important substantive issues are approached only in their accidental aspects. Many puzzling and controversial problems of modern physics appear in this framework as interpretive conventions adopted exclusively for pragmatic reasons. For instance, the Heisenberg's uncertainty principle, is presented as a pragmatic device introduced merely to bar 'the asking of awkward . . . questions.'[33] Electrons turn out to be for Ruse simply 'funny entities. . . which (apparently) have contradictory properties.'[34]

One can, certainly, develop such an approach for the same reason as one can interpret the history of mankind by collecting anecdotes and jokes. A dangerous illusion would arise, however, if one thinks that his collection of entertaining gags may be treated on an equal footing with scientific analysis because our sense of humor is also biologically conditioned and evolutionarily useful.

ORWELLIAN SOCIETIES AND RUSE'S EPISTEMOLOGY

Epistemological principles proposed by Ruse and Wilson remain noncontroversial when applied to the early hominids for whom theoretical knowledge really facilitated their struggle for existence. The same principles result, nonetheless, in grotesque effects if one tries to use them in order to explain the rise of the non-Euclidean geometries or quantum mechanics. The Rusean axiom of the illusory nature of scientific objectivity[35] ultimately implies a denial of the correspondence model proposed in the initial part of this paper. Biological, intellectual and evolutionary factors cannot be regarded as mutually irreducible elements of the process of cognition because the intellectual mixture of necessary facts and epic myths is both genetically determined and subordinate to the criterion of evolutionary advantages. Consequently, evolutionary advantage seems to constitute the ultimate criterion of truth, while the mythopoietic component of our theories varies together with the variation of environmental conditions.

The thesis declaring the utility of knowledge in the evolutionary struggle for survival was formulated in the context of the actually known biological realities of the struggle in question. Let us develop a *Gedanken-experiment* in which the terrestrial conditions of evolution are radically different due to invasion by extraterrestrial humanoids representing a technologically dominant civilization. These humanoids, let us call them humanoids B, were expelled from Andromeda where they were defeated in their struggle for survival by humanoids A who accept a counterpart of Aristotelian logic and our ethics of altruism. Humanoids B propagate an aggression-laden ethics of evolutionary class struggle. Their logic, let us call it dialectical logic, contains as a basic axiom the principle of contradiction asserting that A and non A may be simultaneously accepted for pragmatic reasons. In their social philosophy these creatures proclaim a future classless and truthless society in which all leading roles would be played by the humanoids B who would be simply the most equal members in the future society of equal beings. The content of their theories does not seem especially eccentric when one calls into question the principles formulated by the termite intellectuals in the Wilsonian analogy described above. In this invented story of the Orwellian humanoids I use the same interpretive means that are systematically used by Wilson and Ruse in their arguments.

If the Orwellian situation were extended to the future of biological evolution, the humanoids B would constitute a victorious class in the terrestrial struggle for existence, due to their technological supremacy dependent on discoveries unknown to the species *homo sapiens*. Consistently, in accord with the Rusean principles of sociobiological epistemology, their scientific theories, ethical principles and personal convictions should be considered true since they lead to obvious advantages in the terrestrial struggle for survival. The advantages are open not only for the humanoids B themselves but also for those representatives of the human species who praise dialectical logic and enthusiastically support the program of the construction of the truthless society.

The analogy provided appears grotesque in so far as we think of the humanoids B in Orwellian convention, regarding them as immoral propagators of totalitarian practices and of false ideology. Such an appraisal implies extrapolating our moral principles and criteria of truth onto the representatives of the different species which evolved in entirely different physical and biological conditions. If, according to Ruse's contention, all knowledge is 'grounded in biology' and the biological structure of the humanoids' DNA (if any) differs from ours, it is possible that their logic, axiology, ethics, and social philosophy are also radically different from ours. Since these technologically superior humanoids won the struggle for survival with the human species, their theoretical interpretation of the world must be regarded as evolutionary advantageous and true as well. Ethics of intellectual slavery and conformism would in such a situation be the highest evolutionary achievement of *homo sapiens*.

I do not imagine this story to provide a counterexample to the generally understood sociobiological theory of knowledge. It discloses only inconsistencies contained in Ruse's version of sociobiological epistemology. These inconsistencies result from the cavalier ignoring of rational and critical metascientific analyses of the past. An attempt to replace them by vague metaphors and rhetorical ruses could satisfy only those authors whose visionary declarations prefer something more than substantiated arguments. One must then definitely both de-ruse and de-Ruse the sociobiological theory of knowledge to determine how rational content and biological-evolutionary determinants coexist in our knowledge.

Pontifical Academy of Theology
Kraków

NOTES

[*] This paper was first published in *Historia et Theoria Scientarium*, 1 (1991) 59.
[1] M. Ruse, *Taking Darwin Seriously? A Naturalistic Approach to Philosophy*, Basil Blackwell: 1986, 172. In the subsequent notes this book is denoted by the symbol TD.
[2] M. Ruse, 'Introduction', in: *Nature Animated*, ed. M. Ruse, D. Reidel: Dordrecht 1983, 10.
[3] TD, 279.
[4] I use the expression 'radical sociobiology' to denote the version proposed by Wilson in *On Human Nature* and developed in his works in the 1970s. The term 'moderate sociobiology' denotes sociobiological theories developed by the authors who focus upon biological issues and reduce to a minimum ideological extra-scientific comments.
[5] TD, 279.
[6] TD, 170.
[7] TD, 162.
[8] M. Ruse, 'Darwinism and Determinism', *Zygon*, 22 (1987) 423.
[9] A. Zahavi, 'Mate selection – a selection for a handicap,' *J. Theor. Biol.*, 53 (1975) 205. Cf. R. Dawkins, *The Selfish Gene*, Oxford University Press: London 1976, 171f.
[10] 'Introduction', 5.
[11] A. Newell, 'Artificial Intelligence and the Concept of Mind', in: *Computers Models of Thought and Language*, W.H. Freeman and Co.: San Francisco 1973, 45.
[12] E.O. Wilson, *On Human Nature*, Harvard University Press: London 1978, 201.
[13] 'Conversations at Nobel XVIII,' in: *Darwin's Legacy*, ed. C.L. Hamrum, Harper & Row: San Francisco 1983, 118.
[14] 'Conversations at Nobel XVIII,' in: *Darwin's Legacy*, ed. C.L. Hamrum, Harper & Row: San Francisco 1983, 119.
[15] TD, 175.
[16] E.O. Wilson, *On Human Nature*, Harvard University Press: London 1978, 207.
[17] TD, 191.
[18] TD, 192.
[19] TD, 192.
[20] TD, 192.
[21] TD, 194.
[22] TD, 196.
[23] TD, 206.
[24] TD, 206.
[25] TD, 206.
[26] TD, 173.
[27] TD, 170.
[28] TD, 171.
[29] S.G. Shanker, 'Preface' to *Gödel's Theorem in Focus*, Croom Helm: London/New York 1988), p. vii.
[30] TD, 170 note 4.
[31] M. Ruse, 'Darwinism and determinism', *Zygon*, 22 (1987) 423.
[32] M. Ruse, 'Darwinism and determinism', *Zygon*, 22 (1987) 423.
[33] TD, 157.
[34] TD, 154.
[35] TD, 173.

JÓZEF ŻYCIŃSKI

TACIT KNOWING AND THE RATIONALITY OF SCIENCE

1. INTRODUCTION

Recognition of the tacit and personal component of our knowledge resulted in significant revisions in simplified theories of scientific rationality and objectivity. As a result, Polanyi's epistemological contributions are used not only to justify the conclusions that 'absolute objectivity is a Pickwickian notion'[1] and that 'the search for a universal rationality is Utopian',[2] but also to criticise basic principles of traditional rational epistemology. The author of *Personal Knowledge* is often cited next to Thomas Kuhn, Paul Feyerabend and the later Wittgenstein by authors who try to defend the doctrine of the relativism of truth. According to their arguments, the traditional opposition between objective and subjective, rational and irrational, must collapse because all our knowledge is to some extent tacit, and no set of cognitive procedures can unambiguously grasp the meaning dependent upon empathic indwelling of the human subject.

In an even more radical approach, Polanyi's works are cited next to F. Capra's *The Tao of Physics*, R.M. Pirsig's *Zen and the Art of Motorcycle Maintenance*, and the anti-intellectual texts of the proponents of cultural revolution. Their author is supposed to provide substantial premises for mixing together physics and mysticism, to treat science as metaphor and to counteract the 'modern idolatry' that is expressed in mathematical descriptions of 'the world reduced to quantities and the relationships between them.'[3] In this approach, Polanyi's metascientific contribution is regarded as 'most fruitfully resonant with Taoism' and 'Polanyian-Taoist hermeneutic is intrinsically interesting and highly useful.'[4]

After accepting such an assessment of Polanyi's epistemological legacy, one can easily call into question the thesis of the rationality of science and to contend that science is nothing but one of the forms of our culture while its rationality is entirely generated by the reasonableness of culture.[5] When the thesis of the personal component of our knowledge is translated into the language of radical cognitive sociology, one arrives at an untestable version of cognitive relativism. Its supporters argue that 'all knowledge . . . is conjectural' and 'relative to the local situation of the thinkers who produce it.'[6] After

J. Misiek (ed.), The Problem of Rationality in Science and its Philosophy, 253–263.

adopting such a radical framework one should consistently argue that the truth of '2 + 2 = 4' is relative to the personal situation of the person calculating and that the difference between doing science and playing baseball depends mainly on social conventions relative to given scientific paradigms.

It is not my goal to argue in this paper that there is a difference between science and baseball. I would like only to point out that contemporary attempts to use Polanyi's epistemology to call the rationality of science into question are substantively unfounded. If Polanyi himself stresses the point that no 'meaning achieved by men' may be regarded 'to be merely subjective or private' or 'a mere byproduct of the real interactions that go in the world,'[7] it is substantively ungrounded to pay lip service to his epistemology in arguments which suppose that the essential content of meaning depends primarily on extra-rational determinants.

Polanyi's discovery of the epistemological importance of extra-rational components of knowledge played an important role in overcoming the unrealistic vision of science that was proposed by logical positivists. He adequately described the importance of the inexpressible component in knowledge. This component depends on our personal interpretive preferences, on the overrichnes of our perceptual domain and, last but not least, on our metaphysical commitments.[8] The acceptance of the epistemological importance of these elements reveals the simplifications contained in the positivist image of scientific rationality; it does not justify, however, any form of metascientific irrationalism.

My opinion is that in his epistemology Polanyi remains close to the current of critical rationalism that in a different manner is defended by Popper. Recognition of the tacit component of our knowledge results in a more critical and realistic theory of scientific rationality but never justifies scientific relativism. In spite of unavoidable differences generated by tacit knowing, we can still defend the Popperian thesis of verisimilitude[9] and regard the problem-solving effectiveness of science as an objective hallmark of scientific rationality. In such a framework, in spite of well-known difficulties concerning the criteria of verisimilitude, we may regard scientific research as a gradual approximation to the truth. By interpreting philosophically in the next paragraph the notion of physical Lorentz invariants, I will contend that the same objective truths can be formulated in various conceptual systems where disparate roles are played by tacit and personal factors.

2. MEANING VARIANCE AND CONTENT INVARIANTS

All articulated knowledge is grounded in tacit unarticulated knowing. This very fact makes it possible to claim that essential differences at the level of unarticulated principles result in radical meaning variance of theoretical terms used in rational arguments. The thesis of radical meaning variance was extensively discussed in the debates over the incommensurability of theories separated by scientific revolutions.[10] Contemporary followers of the early Kuhn defend versions of paradigmatic relativism much more radical than the version proposed by Kuhn himself. Whereas the later Kuhn emphasizes that the possible incommensurability of theories does not imply impossibility to compare their content,[11] his radical champions propose much more drastic epistemological assumptions. According to B. Barnes, the socially determined tacit component has to play the decisive role in science, because 'nothing in the nature of things, or the nature of language, or the nature of past usage, determines how we employ, or correctly employ, our terms.'[12]

As a result, the very notion of scientific objectivity loses its traditional meaning[13] when so-called 'objective reality' is regarded as nothing but an output of our creative imagination, dependent upon unarticulated components of our subconsciousness.[14]

It is obvious that such strong epistemological claims are clearly inconsistent with Polanyi's approach in which 'man can transcend his own subjectivity by striving passionately to fulfill his personal obligations to *universal standards*,'[15] standards that are both rational and transsubjective. To avoid simply recapitulating Polanyi's arguments, I would try to justify the cited thesis by providing my own argument based on analysis of epistemological status of Lorentz invariants in relativistic physics.

Contemporary adherents of epistemological relativism in their critiques of the traditional concepts of objective truth and scientific rationality sometimes conceive of Einsteins's relativity theory as a paradigm of all relativistic interpretations. Their arguments, however, seem hardly convincing since it has been a truism in physics that the very expression 'theory of relativity' displays nothing but terminological preferences. There were substantial reasons to call Einstein's theory 'the theory of absoluteness' because it describes both the relative constituents of the physical world (dependent on the adopted frame of reference) and the absolute elements that remain invariants in all systems, irrespective of the linguistic conventions in which they are articulated.[16] The Einsteinian principle of relativity, also called the Lorentz invariance, asserts that physical laws remain the same, irrespective of the adopted inertial system, i.e. they can be considered relative to any inertial system. Consequently, a given statement may be regarded as a law of physics if and only if it is Lorentz invariant in regard to a group of transformations called Poincare group (or the inhomogeneous Lorentz group).[17]

The (Lorentz) invariance of physical laws yields many profound questions that are intriguing both ontologically and epistemologically. When describing philosophically most significant features of the laws of physics, Heinz R. Pagels mentions, among other things, their invariance, universality, simplicity, and susceptibility to mathematical articulation.[18] These features do not depend on the adopted descriptive conventions[19] but constitute objective properties of nature that remain independent of the personal preferences determining our choice of a particular frame of reference.

The features in question provide a difficult interpretative problem for those authors who try to overemphasize the role of the relative, personal and tacit component in our knowledge. It is easy to notice important differences between phenomena investigated in the humanities on the one hand and in the natural sciences on the other. When examining the domain of human culture, an adherent of cultural relativism could call into question the absolute necessity of any cultural form since we can find a human society in which the specific form of customs, rites, or traditions is rejected. The situation appears radically different in the domain of the natural sciences. We cannot accept a 'personal' physics in which, due to the influence of an unarticulated tacit component, laws of physics are not Lorentz invariant. Such a 'physics' would not be physics at all but only a science-fiction play of the imagination.

The inhabitants of Samoa or Borneo can have cultural customs that are as legitimate as European customs, but they cannot have a private physics that would be a real alternative to modern physics. The objective physical reality imposes important constraints on both our creative imagination and our tacit knowing that influences personal biases. There are unsurpassable limits where personal preferences must be subordinate to objective universal principles unless one programmatically champions an irrational theory of knowledge.

We do not know why certain properties of the mathematical description of nature do not depend on the adopted formalism. We do not know either why the language of invariants and symmetries appears so effective in description of the objective reality of physical fields. We can only acknowledge the fact that these objective factors constitute the essence of scientific rationality and belong to the deepest ideas of contemporary physics. They provide – as H. Pagels points out it – 'a symbolic, mathematical language which, when fully grasped, leaves no room for a further reduction in meaning.'[20]

The indicated epistemological regularity might be illustrated by many other similar types of examples that belong either to the natural sciences or to mathematics and logic. All of them depict a philosophically important regularity consisting in the fact that the same substantive content can be expressed by using different linguistic means and formal techniques. This very fact remains epistemologically important because one can no longer defend epistemological relativism by regarding the tacit component of our knowledge as a source of boundless interpretive creativity, in which radically new meaning must be ascribed to all terms due to personal preferences in choosing new linguistic forms and methodological principles. In spite of the unavoidable presence of this component in our intellectual activity, one must acknowledge objective limits to its influence unless our explanatory procedures are to be Orwellian in nature.

3. THE EPISTEMOLOGICAL PRINCIPLE OF UNCERTAINTY

It was Max Born who defined modern science as a result of man's struggle to liberate himself from the subject and its personal experience. The growth of metascientific analysis revealed that the evolution of scientific method proceeded in a direction opposite to that suggested by Born. This evolution categorically confirms what Percy W. Bridgman called 'the first principle of mental dynamics.' The principle in question acknowledges that it is impossible to pursue any scientific discipline whatsoever without accepting personal presuppositions that are adopted either in explicitly articulated form or as a tacit component of knowing.[21]

How could we assess the role of this tacit and personal component if we reject the endeavours that are practiced by radical critics of the objectivity of truth? In my opinion, at the level of metascientific investigations, we discover an epistemological counterpart to the Heisenberg principle of uncertainty discovered in quantum physics in 1927. As Werner Heisenberg pointed out the impossibility of an arbitrarily precise measuring of microphenomena, in a similar manner the epistemological principle of uncertainty discloses the groundlessness of traditional attempts aimed at finding one universally accepted set of cognitive procedures and methodological principles.

Modern metascientific reflections, from Descartes until the logical positivists, was animated by an optimistic faith in the existence of a well-defined set of methodological procedures that would guarantee the certainty of the conclusions achieved. This belief inspired Descartes's *Discourse on Method* as well as Leibniz's search for the *lingua characteristica* and *calculus ratiotinator*, the logical calculus that was supposed to eliminate divergences of opinions in any field of scholarly investigation.

The same optimistic belief can be found in J.S. Mill's works on inductive reasoning and in Carnap's theory of protocol sentences. It was only in 1936 that Hans Reichenbach, in his assessment of Carnap's attempts to determine the unquestionable foundations for the new unified science, compared Carnap to a modern Descartes who searches for 'absolutely certain foundations of science.'[22] This strong declaration seems unrealistic now, 55 years later, when the comments in vogue today deny the existence of *the* scientific method and try to ridicule the neo-Cartesian belief in scientific rationality.

The discovery of insurmountable limitations upon all forms of rational explanation of the world does not imply either methodological anarchism or interpretive relativism. Similar limitations were found in physics by Heisenberg's discovery of the principle of uncertainty. This principle, however, does not justify a total calling into question of the value of the physical interpretation of nature. It only reveals the unacceptability of the most radical ideas that were expressed in Laplace's philosophy of physics. The so-called Laplacean demon was supposed to know with optimal accuracy the parameters of all physical particles in order to predict all future states of the evolving universe on the basis of this knowledge. After Heisenberg's discovery, we realized that the Laplacean demon cannot exist because, due to the quantum effects, in measurement of the so-called conjugate parameters (e.g. energy and time, momentum and positions) an error must appear, at least of the order of Planck's constant ($h = 6.625 \times 10^{-27}$ erg sec).

In spite of the existence of insurmountable barriers in physical studies of microprocesses, quantum physics invariably provides consistent explanations of microphenomena, explanations that are both rational and compatible with empirical data. The discovery of these barriers resulted in important epistemological revisions and in rejection of the naive, over-simplified interpretive schemes; it did not bring the end of physics understood as rational study of objective natural phenomena.

An analogous situation takes place in epistemology. There exists no epistemological counterpart of the Laplacean demon talented enough to free our theories of rationality and scientific objectivity from personal presuppositions and tacit presumptions. The theories and the presuppositions in question form a conjugate system, the unity of which is analogous to the unity of conjugate parameters in quantum mechanics.

We can no longer dream of a one 'only possible' standard of scientific rationality and we should not continue the Cartesian search for the one scientific method that was supposed to guarantee successful discoveries. Various fields of research require different methods and different research procedures. The choice of the latter depends to a certain extent on personal preferences, unarticulated presumptions, and sociologically conditioned determinants. The claim that the latter factors have been more important than the rational components of science not only begs the question but also reveals a personal (tacit, but not necessary) bias toward the irrational stance in epistemology.

Between the Scilla of dogmatic rationalism and the Charybdis of facile relativism we can search for a moderate rationalism in which both Polanyi's and Popper's contributions are necessary so as to work out a critical epistemology liberated from the prejudices that are fashionable at a given period. When pursuing this goal, if we want to free our metascientific reflection from unconfirmed purely visionary ideas as far as possible, it is necessary to take into consideration the status of the most developed scientific disciplines. Although cognitive regularities of the sort that occur in physics need not necessarily take place in the study of cultural or social phenomena, the very fact of the existence of such regularities provides important counterexamples to those metascientific stances in which the notion of scientific objectivity is completely rejected.

It is worthwhile to note that many critics of scientific rationality try systematically to ignore any conclusions provided by the philosophy of physics. To justify such an approach they categorically reject the opinion that 'blissful ignorance of the rational history of sciences'[23] could impede the research of cognitive sociologists who try to explain the content of scientific theories on the basis of socio-cultural determinants. In their assessment, this last opinion is supposed to express nothing but the 'professional imperialism' of the defenders of scientific rationality who adopt a 'surprisingly arrogant and chauvinistic stance with respect to cognitive sociologists.'[24]

4. TACIT PRESUPPOSITIONS OF PHILOSOPHICAL PLURALISM

When in the 18th century Kant developed his critique of metaphysics, he still believed in the certainty and necessity of Newtonian physics. In the same epoch, Voltaire assessed the intellectual situation of the 18th century in a categorical declaration: 'There can be only one geometry and one ethics.' The discovery of non-Euclidean geometries in the next century demonstrated the groundlessness of these optimistic beliefs in the basic unity of our intellectual schemes. The breakdown of the positivist idea of unified science marks the definite end of earlier waiting for a scientist's general agreement on the basis of unshakable metascientific foundations.

When objectively assessing the role of personal components in science, we cannot, at the level of metascientific analysis, await future Newtons who would be able satisfactorily to work out universal methodological principles to guarantee unquestionable conclusions in all domains of scientific research. Polanyi's theory of tacit knowing contributed to a large degree to the discovery of the illusionary nature of similar expectations. This theory was conclusively confirmed by scientific discoveries which revealed that the intrinsic structure of science is much more sophisticated than the structures imagined by champions of simple metascientific schemes. The nature of this structure cannot be determined by taking either rational or extra-rational elements into consideration. Different scientific disciplines exhibit essentially different intrinsic structures, and for this reason no non-trivial epistemology can be universally applied.

Many assertions that seemed iconoclastic when Popper opposed the Vienna Circle and when the first edition of *Personal Knowledge* was published today appear trivial or self-evident. Many themes of long bygone controversies no longer inspire philosophical debates when the former simplifications are critically assessed. After recognizing the failure of the programme in which unified science was supposed to be a purely rational enterprise built on the rock of protocol sentences, we do acknowledge the large spectrum of different epistemological principles and demonstrative procedures. There are domains of science in which deductivist guesses play a positive heuristic role, and domains in which inductivist generalizations are necessary. The complex structure of science can be subordinate neither to the empiricist nor to the deductivist philosophy of science, because in scientific research, rational and empirical components interact with psycho-social determinants in a manner that does not fit any idealized universal principles.

This abandonment of over-simplified views of the nature of scientific knowledge seems to be a very important intellectual attainment of the post-positivist philosophy of science. Its recognition implied by the acknowledgement of the role of personal knowledge does not, however, justify the stance of radical cognitive sociologists who argue that the content of scientific theories is to be determined primarily by extra-rational factors.

In person-laden scientific theories there still remain important content invariants which express the same truths, irrespective of the adopted terminological means and methodological principles. This coexistence of distinct means and invariant truth appears particularly important for the philosophy of science. A long tradition existed in this field, according to which the unity of truth was supposed to be achieved by a programmatic unification of cognitive means. At the present stage of metascientific reflection, we realize that this unity cannot be achieved. The presence of personal components in knowing imposes unsurpassable constraints on our ontological commitments, epistemological choices, methodological predilections. They result in a philosophical pluralism that cannot be *a priori* forbidden.

Contrary to Voltaire's declaration, there exists neither one universally accepted ethics nor one necessary epistemology. Unitarian dreams may easily be regarded as a relic of cognitive totalitarianism. The inevitable pluralism of rival interpretations, however, does not justify the thesis claiming that all these interpretations have equal value. The very existence of Nazi anthropology, Velikovsky's astronomy, or Lysenko's most progressive biology, does not imply that all these pathological interpretations are as good as their scientific counterparts.

We can easily refer to the objective scientific evidence that allows a rationally grounded rejection of these pseudo-explanations. The tacit component of our knowing thus introduces into metascientific analysis an interpretive pluralism analogous to that allowed by Heisenberg's principle in quantum mechanics.

Recognition of the role of the personal constituent of science implies rejection of the Leibnizian dreams of a machine rigorously selecting true and false conclusions. Such a machine, due to the use of logical algorithms and formalized linguistic techniques, was supposed to bring a radical resolution to both philosophers' and scientists' disagreements. Thanks to formalized mechanical reasoning based upon a new universally accepted method of research, the community of scholars was supposed to solve all metaphysical problems within two years and to prepare an encyclopedia of all unified knowledge in five years.[25] The impossibility to complete Leibniz's programme does not imply replacing rigorous methodology by whimsical poetry. It only reveals that the search for truth appears to be a much more complicated enterprise than was thought by proponents of the unsophisticated methodology of the past.

The failure of the simple and ordered methodology may be compared to the collapse of the Aristotelian physics of the perfect superlunar world that was supposed to consist of ideal objects and perfect structures. Such a world does not exist; nonetheless, our physics of imperfect structures remains still an exciting undertaking. The breakdown of the Aristotelian simplifications marked the new intellectual epoch of modern science. An epistemology in which the positivist concept of rationality and objectivity collapsed opens a new period of the critical philosophy of science. Our critical assessment of the heterogeneous components of knowledge introduces important changes into former theories of scientific rationality but still permits us to conceive of scholarly research as a critical search for objective truth.

Pontifical Academy of Theology
Kraków

NOTES

[1] J. Gill, *On Knowing God*, The Westminster Press: Philadelphia 1981, 97.

[2] I. Lakatos, 'Why did Copernicus's Research Programme Supersede Ptolemy's?' in: *Philosophical Papers*, vol. 1, ed. J. Worrall, G. Currie, Cambridge University Press: Cambridge 1980, 176.

[3] R. Jones, *Physics as Metaphor*, Meridian: New York 1982, 14.

[4] J.W. Stines, 'I am the Way: Michael Polanyi's Taoism', *Zygon*, 20 (1985) 59, 62.

[5] Cf. B. Barnes, 'Problems of Intelligibility and Paradigm Instances', in: *Scientific Rationality: The Sociological Turn*, ed. J.R. Brown, Reidel: Dordrecht 1984, 113.

[6] D. Bloor, *Knowledge and Social Imagery*, Routledge & Kegan Paul: London 1976, 142.

[7] M. Polanyi, H. Prosch, *Meaning*, University of Chicago Press: Chicago 1975, 182. This declaration is directed not against the Strong Programme of the School of Edinburgh, but – as the authors clarify in the context – against Marxist epistemology.

[8] Cf. T. Settle, 'Is Scientific Knowledge Rationally Justified?' in: *Basic Issues in the Philosophy of Science*, New York 1976, 31.

[9] The thesis, in W.H. Newton-Smith's wording, asserts that in science we 'would have progress after all, for our theories would be capturing more and more truth about the world.' Consistently, 'rationality consists in believing in those theories which it is most reasonable to presume have the highest degree of verisimilitude among the available rival theories.' *The Rationality of Science*, Routledge & Kegan Paul: London 1981, 184.

[10] The very expression 'the radical meaning variance' was introduced by C.R. Kordig, *The Justification of Scientific Change*, Reidel: Dordrecht 1971.

[11] Cf. for instance, his declaration: 'In applying the term 'incommensurability' to theories, I had intended only to insist that there was no common language within which both could be fully expressed and which could therefore be used in a point-by-point comparison between them.' 'Theory Change as Structure Change', *Erkenntnis*, 10 (1976) 190.

[12] B. Barnes, *T.S. Kuhn and Social Science*, Routledge & Kegan Paul: London 1982, 30.

[13] Cf. A. Gouldner, *The Coming Crisis of Western Sociology*, New York 1970, 103, where he asserts categorically that 'objectivity is the compensation men offer themselves when their capacity to love has been crippled.'

[14] Cf. for instance, R. Sukenick, 'Upward and Juanward: The Possible Dream,' in: *Seeing Castaneda: Reactions to the 'Don Juan' Writings of Carlos Castaneda*, New York 1976, 113.

[15] M. Polanyi, *Personal Knowledge: Towards a Post-Critical Philosophy*, The University of Chicago Press: Chicago 1962, 17. Emphasis mine – JZ.

[16] Cf. J. Rayski, *Foundations of Modern Theoretical Physics*, PWN: Warszawa-Kraków 1983, 18.

[17] A more extensive exposition of this problem may be found in M. Heller, 'A Comment on Lorentz Invariance', *Philosophy in Science*, 1 (1983) 90f.

[18] H.R. Pagels, *The Cosmic Code: Quantum Physics as the Language of Nature*, Bantam Books: London 1983, 292.

[19] Unless, of course, private non-standard patterns of simplicity or universality are adopted.

[20] H. Pagels, *Perfect Symmetry: The Search for the Beginning of Time*, Bantam Books: London 1986, 191.

[21] The status of the Bridgmanian principles is analysed in J. Życiński, *The Structure of the Metascientific Revolution*, Pachart: Tucson 1988, 137–145.

[22] H. Reichenbach, 'Logistic Empiricism in Germany and the Present State of its Problems,' *J. of Phil.*, 33 (1936) 140–160.

[23] The expression used by L. Laudan in his critique of K. Mannheim in *Progress and Its Problems*, Routledge & Kegan Paul: London 1977, 209.

[24] R.C. Jennings, 'Truth, Rationality and the Sociology of Science,' *Brit. J. Phil. Sc.*, 35 (1984) 209f.

[25] *Die Philosophischen Schriften von Gottfried Wilhelm Leibniz*, ed. C.I. Gerhardt, George Olms: Hildesheim 1890, vol. 7, 187.

INDEX OF NAMES

INDEX OF SUBJECTS

Boston Studies in the Philosophy of Science

105. F. Burwick (ed.): *Approaches to Organic Form.* Permutations in Science and Culture. 1987 ISBN 90-277-2541-1
106. M. Almási: *The Philosophy of Appearances.* Translated from Hungarian. 1989 ISBN 90-277-2150-5
107. S. Hook, W.L. O'Neill and R. O'Toole (eds.): *Philosophy, History and Social Action.* Essays in Honor of Lewis Feuer. With an Autobiographical Essay by L. Feuer. 1988 ISBN 90-277-2644-2
108. I. Hronszky, M. Fehér and B. Dajka: *Scientific Knowledge Socialized.* Selected Proceedings of the 5th Joint International Conference on the History and Philosophy of Science organized by the IUHPS (Veszprém, Hungary, 1984). 1988 ISBN 90-277-2284-6
109. P. Tillers and E.D. Green (eds.): *Probability and Inference in the Law of Evidence.* The Uses and Limits of Bayesianism. 1988 ISBN 90-277-2689-2
110. E. Ullmann-Margalit (ed.): *Science in Reflection.* The Israel Colloquium: Studies in History, Philosophy, and Sociology of Science, Vol. 3. 1988 ISBN 90-277-2712-0; Pb 90-277-2713-9
111. K. Gavroglu, Y. Goudaroulis and P. Nicolacopoulos (eds.): *Imre Lakatos and Theories of Scientific Change.* 1989 ISBN 90-277-2766-X
112. B. Glassner and J.D. Moreno (eds.): *The Qualitative-Quantitative Distinction in the Social Sciences.* 1989 ISBN 90-277-2829-1
113. K. Arens: *Structures of Knowing.* Psychologies of the 19th Century. 1989 ISBN 0-7923-0009-2
114. A. Janik: *Style, Politics and the Future of Philosophy.* 1989 ISBN 0-7923-0056-4
115. F. Amrine (ed.): *Literature and Science as Modes of Expression.* With an Introduction by S. Weininger. 1989 ISBN 0-7923-0133-1
116. J.R. Brown and J. Mittelstrass (eds.): *An Intimate Relation.* Studies in the History and Philosophy of Science. Presented to Robert E. Butts on His 60th Birthday. 1989 ISBN 0-7923-0169-2
117. F. D'Agostino and I.C. Jarvie (eds.): *Freedom and Rationality.* Essays in Honor of John Watkins. 1989 ISBN 0-7923-0264-8
118. D. Zolo: *Reflexive Epistemology.* The Philosophical Legacy of Otto Neurath. 1989 ISBN 0-7923-0320-2
119. M. Kearn, B.S. Philips and R.S. Cohen (eds.): *Georg Simmel and Contemporary Sociology.* 1989 ISBN 0-7923-0407-1
120. T.H. Levere and W.R. Shea (eds.): *Nature, Experiment and the Science.* Essays on Galileo and the Nature of Science. In Honour of Stillman Drake. 1989 ISBN 0-7923-0420-9
121. P. Nicolacopoulos (ed.): *Greek Studies in the Philosophy and History of Science.* 1990 ISBN 0-7923-0717-8
122. R. Cooke and D. Costantini (eds.): *Statistics in Science.* The Foundations of Statistical Methods in Biology, Physics and Economics. 1990 ISBN 0-7923-0797-6

Boston Studies in the Philosophy of Science

123. P. Duhem: *The Origins of Statics.* Translated from French by G.F. Leneaux, V.N. Vagliente and G.H. Wagner. With an Introduction by S.L. Jaki. 1991 ISBN 0-7923-0898-0
124. H. Kamerlingh Onnes: *Through Measurement to Knowledge.* The Selected Papers, 1853-1926. Edited and with an Introduction by K. Gavroglu and Y. Goudaroulis. 1991 ISBN 0-7923-0825-5
125. M. Čapek: *The New Aspects of Time: Its Continuity and Novelties.* Selected Papers in the Philosophy of Science. 1991 ISBN 0-7923-0911-1
126. S. Unguru (ed.): *Physics, Cosmology and Astronomy, 1300-1700.* Tension and Accommodation. 1991 ISBN 0-7923-1022-5
127. Z. Bechler: *Newton's Physics on the Conceptual Structure of the Scientific Revolution.* 1991 ISBN 0-7923-1054-3
128. É. Meyerson: *Explanation in the Sciences.* Translated from French by M-A. Siple and D.A. Siple. 1991 ISBN 0-7923-1129-9
129. A.I. Tauber (ed.): *Organism and the Origins of Self.* 1991 ISBN 0-7923-1185-X
130. F.J. Varela and J-P. Dupuy (eds.): *Understanding Origins.* Contemporary Views on the Origin of Life, Mind and Society. 1992 ISBN 0-7923-1251-1
131. G.L. Pandit: *Methodological Variance.* Essays in Epistemological Ontology and the Methodology of Science. 1991 ISBN 0-7923-1263-5
132. G. Munévar (ed.): *Beyond Reason.* Essays on the Philosophy of Paul Feyerabend. 1991 ISBN 0-7923-1272-4
133. T.E. Uebel (ed.): *Rediscovering the Forgotten Vienna Circle.* Austrian Studies on Otto Neurath and the Vienna Circle. Partly translated from German. 1991 ISBN 0-7923-1276-7
134. W.R. Woodward and R.S. Cohen (eds.): *World Views and Scientific Discipline Formation.* Science Studies in the [former] German Democratic Republic. Partly translated from German by W.R. Woodward. 1991 ISBN 0-7923-1286-4
135. P. Zambelli: *The Speculum Astronomiae and Its Enigma.* Astrology, Theology and Science in Albertus Magnus and His Contemporaries. 1992 ISBN 0-7923-1380-1
136. P. Petitjean, C. Jami and A.M. Moulin (eds.): *Science and Empires.* Historical Studies about Scientific Development and European Expansion. ISBN 0-7923-1518-9
137. W.A. Wallace: *Galileo's Logic of Discovery and Proof.* The Background, Content, and Use of His Appropriated Treatises on Aristotle's *Posterior Analytics.* 1992 ISBN 0-7923-1577-4
138. W.A. Wallace: *Galileo's Logical Treatises.* A Translation, with Notes and Commentary, of His Appropriated Latin Questions on Aristotle's *Posterior Analytics.* 1992 ISBN 0-7923-1578-2
Set (137 + 138) ISBN 0-7923-1579-0

Boston Studies in the Philosophy of Science

139. M.J. Nye, J.L. Richards and R.H. Stuewer (eds.): *The Invention of Physical Science.* Intersections of Mathematics, Theology and Natural Philosophy since the Seventeenth Century. Essays in Honor of Erwin N. Hiebert. 1992 ISBN 0-7923-1753-X
140. G. Corsi, M.L. dalla Chiara and G.C. Ghirardi (eds.): *Bridging the Gap: Philosophy, Mathematics and Physics.* Lectures on the Foundations of Science. 1992 ISBN 0-7923-1761-0
141. C.-H. Lin and D. Fu (eds.): *Philosophy and Conceptual History of Science in Taiwan.* 1992 ISBN 0-7923-1766-1
142. S. Sarkar (ed.): *The Founders of Evolutionary Genetics.* A Centenary Reappraisal. 1992 ISBN 0-7923-1777-7
143. J. Blackmore (ed.): *Ernst Mach – A Deeper Look.* Documents and New Perspectives. 1992 ISBN 0-7923-1853-6
144. P. Kroes and M. Bakker (eds.): *Technological Development and Science in the Industrial Age.* New Perspectives on the Science–Technology Relationship. 1992 ISBN 0-7923-1898-6
145. S. Amsterdamski: *Between History and Method.* Disputes about the Rationality of Science. 1992 ISBN 0-7923-1941-9
146. E. Ullmann-Margalit (ed.): *The Scientific Enterprise.* The Bar-Hillel Colloquium: Studies in History, Philosophy, and Sociology of Science, Volume 4. 1992 ISBN 0-7923-1992-3
147. L. Embree (ed.): *Metaarchaeology.* Reflections by Archaeologists and Philosophers. 1992 ISBN 0-7923-2023-9
148. S. French and H. Kamminga (eds.): *Correspondence, Invariance and Heuristics.* Essays in Honour of Heinz Post. 1993 ISBN 0-7923-2085-9
149. M. Bunzl: *The Context of Explanation.* 1993 ISBN 0-7923-2153-7
150. I.B. Cohen (ed.): *The Natural Sciences and the Social Sciences.* Some Critical and Historical Perspectives. 1994 ISBN 0-7923-2223-1
151. K. Gavroglu, Y. Christianidis and E. Nicolaidis (eds.): *Trends in the Historiography of Science.* 1994 ISBN 0-7923-2255-X
152. S. Poggi and M. Bossi (eds.): *Romanticism in Science.* Science in Europe, 1790–1840. 1994 ISBN 0-7923-2336-X
153. J. Faye and H.J. Folse (eds.): *Niels Bohr and Contemporary Philosophy.* 1994 ISBN 0-7923-2378-5
154. C.C. Gould and R.S. Cohen (eds.): *Artifacts, Representations, and Social Practice.* Essays for Marx W. Wartofsky. 1994 ISBN 0-7923-2481-1
155. R.E. Butts: *Historical Pragmatics.* Philosophical Essays. 1993 ISBN 0-7923-2498-6
156. R. Rashed: *The Development of Arabic Mathematics: Between Arithmetic and Algebra.* Translated from French by A.F.W. Armstrong. 1994 ISBN 0-7923-2565-6

Boston Studies in the Philosophy of Science

157. I. Szumilewicz-Lachman (ed.): *Zygmunt Zawirski: His Life and Work.* With Selected Writings on Time, Logic and the Methodology of Science. Translations by Feliks Lachman. Ed. by R.S. Cohen, with the assistance of B. Bergo. 1994 ISBN 0-7923-2566-4
158. S.N. Haq: *Names, Natures and Things.* The Alchemist Jābir ibn Ḥayyān and His *Kitāb al-Aḥjār* (Book of Stones). 1994 ISBN 0-7923-2587-7
159. P. Plaass: *Kant's Theory of Natural Science.* Translation, Analytic Introduction and Commentary by Alfred E. and Maria G. Miller. 1994 ISBN 0-7923-2750-0
160. J. Misiek (ed.): *The Problem of Rationality in Science and its Philosophy.* On Popper vs. Polanyi. The Polish Conferences 1988–89. 1995 ISBN 0-7923-2925-2
161. I.C. Jarvie and N. Laor (eds.): *Critical Rationalism, Metaphysics and Science.* Essays for Joseph Agassi, Volume I. 1995 ISBN 0-7923-2960-0
162. I.C. Jarvie and N. Laor (eds.): *Critical Rationalism, the Social Sciences and the Humanities.* Essays for Joseph Agassi, Volume II. 1995 ISBN 0-7923-2961-9
     Set (161–162) ISBN 0-7923-2962-7
163. K. Gavroglu, J. Stachel and M.W. Wartofsky (eds.): *Physics, Philosophy, and the Scientific Community.* Essays in the Philosophy and History of the Natural Sciences and Mathematics. In Honor of Robert S. Cohen. 1995 ISBN 0-7923-2988-0
164. K. Gavroglu, J. Stachel and M.W. Wartofsky (eds.): *Science, Politics and Social Practice.* Essays on Marxism and Science, Philosophy of Culture and the Social Sciences. In Honor of Robert S. Cohen. 1995 ISBN 0-7923-2989-9
165. K. Gavroglu, J. Stachel and M.W. Wartofsky (eds.): *Science, Mind and Art.* Essays on Science and the Humanistic Understanding in Art, Epistemology, Religion and Ethics. Essays in Honor of Robert S. Cohen. 1995 ISBN 0-7923-2990-2
     Set (163–165) ISBN 0-7923-2991-0

Also of interest:

R.S. Cohen and M.W. Wartofsky (eds.): *A Portrait of Twenty-Five Years Boston Colloquia for the Philosophy of Science, 1960-1985.* 1985 ISBN Pb 90-277-1971-3

Previous volumes are still available.

KLUWER ACADEMIC PUBLISHERS – DORDRECHT / BOSTON / LONDON

GPSR Compliance

The European Union's (EU) General Product Safety Regulation (GPSR) is a set of rules that requires consumer products to be safe and our obligations to ensure this.

If you have any concerns about our products, you can contact us on ProductSafety@springernature.com

In case Publisher is established outside the EU, the EU authorized representative is:

Springer Nature Customer Service Center GmbH
Europaplatz 3
69115 Heidelberg, Germany

Batch number: 10755292

Printed by Printforce, the Netherlands